The Discovery of Radium

Research on Radioactive Substances

Marie Skłodowska Curie

Quaternion Books

Part I published in 1921 as *The Discovery of Radium*, Ellen S. Richards
Monographs No. 2, by Vassar College, New York, USA.
Part II published in 1904 as *Recherches sur les substances radioactives
(Deuxième édition, revue et corrigée)* by Gauthier-Villars, France.

This edition first published in 2020 by Quaternion.
© 2020 Quaternion
www.quaternionbooks.com

ISBN 9798635640418

Sourced from the public domain, this work has been translated
and curated by Alessandro Patruno.

Contents

Part I.

The Discovery of Radium

Address by Madame M. Curie at Vassar College
May 14, 1921

Prefactory Note

In her recent visit to America, Madame Curie conferred a special honor upon Vassar College by delivering in the chapel on the evening of May fourteenth the only extended address which she made in this country. In a simple, straightforward way she told the story of her great achievement. One realized how, closely environed by all the great realities of human experience, in the face of tremendous difficulties and with limited resources, she had pursued undaunted her search for truth.

The discovery of radium gave Madame Curie immediate distinction among scientists on account of the extremely significant contribution she thereby made to the great ultimate problem of physical science, the constitution of matter. The striking properties possessed by radium gave to its discovery a world-wide interest, all the more intense because of the hope which was inspired by the possible healing qualities of the radiations from this new element.

That hope is being realized in large measure. It is therefore fitting that this address should have been given by Madame Curie at Vassar and that it should now be circulated among the members of the college under the foundation in memory of Ellen S. Richards, who devoted her life to the public health.

Edna Carter
Chairman of the Department of Physics.

The Discovery of Radium

I could tell you many things about radium and radioactivity and it would take a long time. But as we can not do that, I shall only give you a short account of my early work about radium. Radium is no more a baby, it is more than twenty years old, but the conditions of the discovery were somewhat peculiar, and so it is always of interest to remember them and to explain them.

We must go back to the year 1897. Professor Curie and I worked at that time in the laboratory of the school of Physics and Chemistry where Professor Curie held his lectures. I was engaged in some work on uranium rays which had been discovered two years before by Professor Becquerel. I shall tell you how these uranium rays may be detected. If you take a photographic plate and wrap it in black paper and then on this plate, protected from ordinary light, put some uranium salt and leave it a day, and the next day the plate is developed, you notice on the plate a black spot at the place where the uranium salt was. This spot has been made by special rays which are given out by the uranium and are able to make an impression on the plate in the same way as ordinary light. You can also test those rays in another way, by placing them on an electroscope. You know what an electroscope is. If you charge it, you can keep it charged several hours and more, unless uranium salts are placed near to it. But if this is the case the electroscope loses its charge and the gold or aluminum leaf falls gradually in a progressive way. The speed with which the leaf moves may be used as a measure of the intensity of the rays; the greater the speed, the greater the intensity.

I spent some time in studying the way of making good measurements of the uranium rays, and then I wanted to know if there were other elements, giving out rays of the same kind. So I took up a work about all known elements, and their compounds and found that uranium compounds are active and also all thorium compounds, but other elements were not found active, nor were their compounds. As for the uranium and thorium compounds, I found that they were active in proportion to their uranium or thorium content. The more uranium or thorium, the greater the activity, the activity being an atomic property of the elements, uranium and thorium.

Then I took up measurements of minerals and I found that several of those which contain uranium or thorium or both were active. But then the activity was not what I could expect, it was greater than for uranium or thorium compounds like the oxides which are almost entirely composed of these elements. Then I thought that there should be in the minerals some unknown element having a much greater radioactivity than uranium or thorium. And I wanted to find and to separate that element, and I settled to that work with Professor Curie. We thought it would be done in several weeks or months, but it was not so. It took many years of hard work to finish that task. There was not one new element, there were several of them. But the most important is radium which could be separated in a pure state.

All the tests for the separation were done by the method of electrical measurements with some kind of electroscope. We just had to make chemical separations and to examine all products obtained with respect to their activity. The product which retained the radioactivity was considered as that one which had kept the new element; and, as the radioactivity was more strong in some products, we knew that we had succeeded in concentrating the new element. The radioactivity was used in the same way as a spectroscopical test.

The difficulty was that there is not much radium in a mineral; this we did not know at the beginning. But we now know that there is not even one part of radium in a million parts of good ore. And too, to get a small quantity of pure radium salt, one is obliged to work up a huge quantity of ore. And that was very hard in a laboratory.

We had not even a good laboratory at that time. We worked in a hangar where there were no improvements, no good chemical arrangements. We had no help, no money. And because of that the work could not go on as it would have done under better conditions. I did myself the numerous crystalizations which were wanted to get the radium salt separated from the barium salt with which it is obtained out of the ore. And in 1902 I finally succeeded in getting pure radium chloride and determining the atomic weight of the new element radium, which is 226 while that of barium is only 137.

Later I could also separate the metal radium, but that was a very difficult work; and, as it is not necessary for the use of radium to have it in this state, it is not generally prepared that way.

Now, the special interest of radium is in the intensity of its rays which is several million times greater than the uranium rays. And the effects of the rays make the radium so important. If we take a practical point of view, then the most important property of the rays is the production of physiological effects on the cells of the human organism. These effects may be used for the cure of several diseases. Good results have been obtained in many cases. What is considered particularly important is the treatment of cancer. The medical utilization of radium makes it necessary to get that element in sufficient quantities. And so a factory of radium was started to begin with in France, and later in America where a big quantity of ore named carnotite is available. America does produce many grams of radium every year but the price is still very high because the quantity of radium contained in the ore is so small. The radium is more than a hundred thousand times dearer than gold.

But we must not forget that when radium was discovered no one knew that it would prove useful in hospitals. The work was one of pure science. And this is a proof that scientific work must not be considered from the point of view of the direct usefulness of it. It must be done for itself, for the beauty of science, and then there is always the chance that a scientific discovery may become like the radium a benefit for humanity.

But science is not rich, it does not dispose of important means, it does not generally meet recognition before the material usefulness of it has been proved. The factories produce many grams of radium every year, but the laboratories have very small quantities. It is the same for my laboratory and I am very grateful to the American women who wish me to have more of radium and give me the opportunity of doing more work with it.

The scientific history of radium is beautiful. The properties of the rays have been studied very closely. We know that particles are expelled from radium with a very great velocity near to that of the light. We know that the atoms of radium are destroyed by expulsion of these particles, some of which are atoms of helium. And in that way it has been proved that the radioactive elements are constantly disintegrating and that they produce at the end ordinary elements, principally helium and lead. That is, as you see, a theory of transformation of atoms which are not stable, as was believed before, but may undergo spontaneous changes.

Radium is not alone in having these properties. Many having other radioelements are known already, the polonium, the mesothorium, the radiothorium, the actinium. We know also radioactive gases, named emanations. There is a great variety of substances and effects in radioactivity. There is always a vast field left to experimentation and I hope that we may have some beautiful progress in the following years. It is my earnest desire that some of you should carry on this scientific work and keep for your ambition the determination to make a permanent contribution to science.

M. Curie.

Part II.

Research on Radioactive Substances

Presented to the Faculty of Sciences at the
Université de la Sorbonne, Paris

Introduction

The purpose of this work is to present the research I have been doing for more than four years on radioactive substances. I started this research with a study of the uranium radiation, discovered by M. Becquerel. The results to which this work led me seemed to open up such an interesting path, that M. Curie[1] abandoned his work on which he was engaged, and joined me in the effort of extracting the new radioactive substances and continue the study of their properties.

Since the beginning of our research we believed we should lend samples of the substances, discovered and prepared by us, to some physicists, first of all to M. Becquerel, who is responsible for the discovery of the uranium rays. In this way we have made it easier for ourselves to research new radioactive substances. Following our first publications, M. Giesel, in Germany, also began to prepare these substances and passed on samples to several German scientists. Finally, these substances were put on sale in France and Germany, and the subject growing in importance gave rise to a scientific movement, such that numerous memoirs have appeared, and are constantly appearing on radioactive substances, principally abroad. The results of the various French and foreign works are necessarily entangled, as is the case with any new subject of study in course of investigation. The face of the question is

---

[1]The French courtesy title "monsieur" (plural "messieurs") is abbreviated in "M." (plural "MM."), equivalent to the English "mister". M. Curie here refers to Pierre Curie.

changing, so to speak, from day to day.

However, from the chemical point of view, a point is definitely established: — i.e., the existence of a new highly radioactive element: radium. Preparing pure radium chloride and the determining the atomic weight of radium is the most important part of my personal work. At the same time, whilst this work adds to the elements currently known a new element with very curious properties, a new method of chemical research is found and justified. This method, based on radioactivity, considered as an atomic property of matter, is precisely the one that enabled us, M. Curie and myself, to discover the existence of radium.

If, from the chemical point of view, the question we originally asked ourselves can be considered as resolved, the study of the physical properties of the radioactive substances keeps evolving. Certain important points have been established, but a large number of conclusions are still provisional. This is not surprising, considering the complexity of the phenomena that give rise to radioactivity and the differences that exist between the various radioactive substances. The researches of the various physicists who study these substances constantly meet and overlap. While seeking to conform to the precise purpose of this work and to expose my own research, I was obliged at the same time to mention results of other work whose knowledge is essential.

I also desired to make this work an overview of the current state of the question.

I carried out this Work in the laboratories of the School of Industrial Physics and Chemistry in Paris, with the permission of Schützenberger, late Director of the School, and M. Lauth, the current Director. I would like to express here my gratitude for the kind hospitality I have received in this School.

Historical.

The discovery of the phenomena of radioactivity is connected with researches carried out since the discovery of the Röntgen rays on the photographic effects of phosphorescent and fluorescent substances.

The first tubes producing Röntgen rays were tubes without the metallic anti-cathode. The Röntgen ray source was on the glass wall struck by the cathode rays; at the same time this wall was strongly fluorescent. The question then was whether the emission of Röntgen rays necessarily accompanied the production of fluorescence, whatever the cause of the latter. This idea was first stated by M. Henri Poincaré[2].

Shortly thereafter, M. Henry announced that he had obtained photographic prints through black paper using phosphorescent zinc sulphide[3]. M. Niewenglowski obtained the same phenomenon with calcium sulphide exposed to the light[4]. Finally, M. Troost obtained strong photographic impressions with zinc sulphide artificially phosphorescent acting through black paper and a thick cardboard[5].

The experiences which have been just cited could not be reproduced, despite the numerous attempts made for this purpose. It can therefore in no way be taken as proven that zinc sulphide and calcium sulphide are capable of emitting, under the action of light, invisible radiation which passes through black paper and acts on photographic plates.

M. Becquerel has carried out similar experiments on uranium

---

[2] *Revue générale des Sciences*, 30 January 1896.
[3] *Comptes rendus*, Vol. CXXII, p. 312
[4] *Comptes rendus*, Vol. CXXII, p. 386.
[5] *Comptes rendus*, Vol. CXXII, p. 564.

salts, some of which are fluorescent[6]. He obtained photographic impressions through black paper with double uranyl and potassium sulphate.

M. Becquerel first believed that this salt, which is fluorescent, behaved like zinc sulfide and calcium sulfide in the experiments of MM. Henry, Niewenglowski, and Troost. But the rest of his experiments showed that the phenomenon observed was in no way related to fluorescence. It is not necessary that the salt is fluorescent; furthermore, uranium and all its compounds, fluorescent or not, act in the same manner, and metallic uranium is the most active. M. Becquerel then found that by placing uranium compounds in complete darkness, they continue to impress photographic plates through black paper for years. M. Becquerel assumed that uranium and its compounds emit specific rays — uranium rays. He proved that these rays can penetrate thin metallic screens, and that they discharge electrified bodies. He also made experiments from which he concluded that uranium rays experience reflection, refraction, and polarisation.

The work of other physicists (Elster and Geitel, Lord Kelvin, Schmidt, Rutherford, Beattie, and Smoluchowski) confirms and extends the results of the research of M. Becquerel, except with regard to reflection, refraction, and polarisation of uranium rays, which, from this point of view, behave like the Röntgen rays, as was recognised by Mr. Rutherford first and then by M. Becquerel himself.

---

[6]Becquerel, *Comptes rendus*, 1896 (several Notes).

Chapter I. Radioactivity of Uranium and Thorium. Radioactive Minerals.

Becquerel Rays. — The uranium rays discovered by M. Becquerel impress photographic plates screened from light; they can penetrate all solid, liquid, and gaseous substances, provided that the thickness is sufficiently thin; in passing through gases, they cause them to become weakly conductive of electricity[1].

These properties of the uranium compounds are not due to any known excitatory cause. The radiation seems spontaneous; it does not decrease in intensity, even when the uranium compounds are kept in complete darkness for years; it is therefore not a particular phosphorescence produced by light.

The spontaneity and persistence of the uranium radiation appear as a completely extraordinary physical phenomenon. M. Becquerel kept a piece of uranium for several years in the dark, and found that after that time the action upon the photographic plate had not changed significantly. MM. Elster and Geitel made a similar experiment and also found that the action was constant[2].

I measured the intensity of radiation of uranium using the effect of this radiation on the conductivity of air. The measurement method will be explained later. I thus obtained numbers which prove the persistence of radiation within the limits of accuracy

---

[1]Becquerel, *Comptes rendus*, 1896 (several Notes).
[2]Becquerel, *Comptes rendus*, Vol. CXXVIII, p. 771. — Elster and Geitel, *Beibl.*, Vol. XXI, p. 455.

of the experiments, that is close to 2 or 3 percent[3].

For these measurements a metallic plate was used covered with a layer of powdered uranium; this plate was also not kept in the dark; this condition having appeared unimportant, according to the experimenters cited above. The number of measurements taken with this plate is very large, and they actually extend over a period of five years.

Research was done to determine whether other substances can act like uranium compounds. M. Schmidt was the first to publish that thorium and its compounds possess exactly the same property[4]. Similar work done at the same time, has given me the same result. I published this work, not yet having knowledge of the publication of M. Schmidt[5].

We shall say that uranium, thorium, and their compounds emit Becquerel rays. I have called radioactive those substances which give rise to such an emission[6]. This name has since been generally adopted.

In their photographic and electric effects, the Becquerel rays are similar to the Röntgen rays. Like the latter, they also have the ability of penetrating all matter. But their penetrating power is extremely different: the rays of uranium and thorium are stopped by a few millimetres of solid matter, and cannot cross in the air a distance greater than a few centimetres; at least this is the case for most of the radiation.

The researches of various physicists, and in the first place of Mr. Rutherford, have shown that the Becquerel rays experience neither regular reflection, nor refraction, nor polarisation[7].

---

[3]Mme. Curie, *Revue générale des Sciences*, January 1899.
[4]Schmidt, *Wied. Ann.*, Vol. LXV, p. 141.
[5]Mme Curie, *Comptes rendus*, April 1898.
[6]P. Curie and Mme Curie, *Comptes rendus*, 18 July 1898.
[7]Rutherford, *Phil. Mag.*, January 1899.

The weak penetrating power of uranium and thorium rays would point to their similarity to the secondary rays which are produced by the Röntgen rays, and whose study was made by M. Sagnac[8], rather than to the Röntgen rays themselves.

For the rest, the Becquerel rays might be classified as cathode rays propagated in the air. It is now known that these different analogies are all legitimate.

Measurement of the Intensity of Radiation.

The method used consists in measuring the conductivity acquired by air under the action of radioactive substances; this method has the advantage of being rapid and provides numbers which are comparable. The device I have used for this purpose consists essentially of a plate condenser, *AB* (Fig. 1).

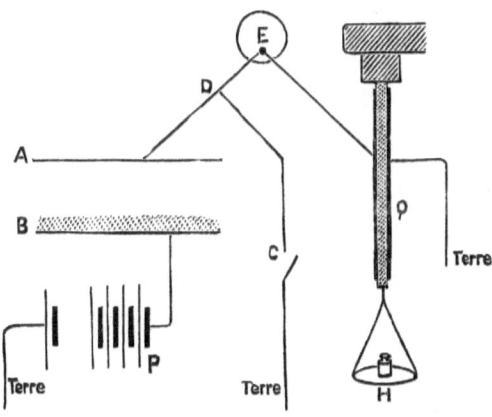

Figure 1.

The finely powdered active substance is spread over the plate *B*, making the air between the plates a conductor. To measure this conductivity, the plate *B* is brought to a high potential by

---

[8]Sagnac, *Comptes rendus*, 1897, 1898, 1899 (several Noies).

connecting it with one of the poles of a battery of small accumulators, P, the other pole being connected to earth. As plate A is kept at the ground potential by the connection CD, an electric current is set up between the two plates. The potential of plate A is recorded by an electrometer, E. If the earth connection is interrupted at C, the plate A charges, and this charge deflects the electrometer. The velocity of the deflection is proportional to the intensity of the current, and can be used to measure the latter.

However, it is preferable to make this measurement by compensating the charge on plate A, so as to cause no deflection of the electrometer. The charges in question are extremely weak; they can be compensated by means of a piezoelectric quartz Q, one sheath of which is connected to plate A and the other to earth. The quart plate is subjected to a known tension, produced by weights placed in a plate, π; the tension is produced progressively, and has the effect of gradually generating a known quantity of electricity during the time that is measured. The operation can be regulated in such a way that, at each instant, there is compensation between the quantity of electricity that crosses the condenser and that of the opposite sign provided by the quartz[9]. We can thus measure *in absolute units* the amount of electricity passing through the condenser for a given time, i.e., the *intensity of the current*. The measurement is independent of the sensitivity of the electrometer.

By carrying out a certain number of measurements of this kind, it is seen that radioactivity is a phenomenon capable of being measured with a certain accuracy. It varies little with tem-

---

[9]This is very easily achieved by supporting the weight by hand and only allowing it to weigh gradually on the plate, this in order to keep the index of the electrometer at zero. With a little bit of practice we get the trick exactly right to succeed in this operation. This method of measuring weak currents was described by M. J. Curie in his thesis.

perature; it is hardly affected by variations in the temperature of the surroundings; it is not influenced by incandescence of the active substance. The intensity of the current flowing through the condenser increases with the surface of the plates. For a given condenser and a given substance the current increases with the potential difference between the plates, with the pressure of the gas which fills the condenser, and with the distance of the plates (provided that this distance is not too great in comparison with the diameter). In every case, for large potential differences, the current tends towards a limiting value, which is practically constant. This is the *saturation current*, or *limiting current*. Likewise, for a certain sufficiently large distance between the plates, the current hardly varies any longer with the distance. It is the current obtained under these conditions that was taken as the measure of radioactivity in my research, the condenser being placed in air at atmospheric pressure.

Here, by way of example, I show curves representing the intensity of the current as a function of the average field established between the plates for two different plate distances. Plate B was covered with a thin layer of powdered metallic uranium; plate A, connected with the electrometer, was provided with a guard-ring.

Fig. 2 shows that the current intensity becomes constant for high potential differences between the plates. Fig. 3 represents the same curves on another scale, and only includes the results relative to the small potential differences. At the origin, the curve is rectilinear; the ratio of the intensity of the current to the difference of potential is constant for low voltages, and represents the initial conductance between the plates. We can therefore distinguish two important characteristic constants of the observed phenomenon: — (1) The initial conductance for small differences of potential; (2) the limiting current for large potential

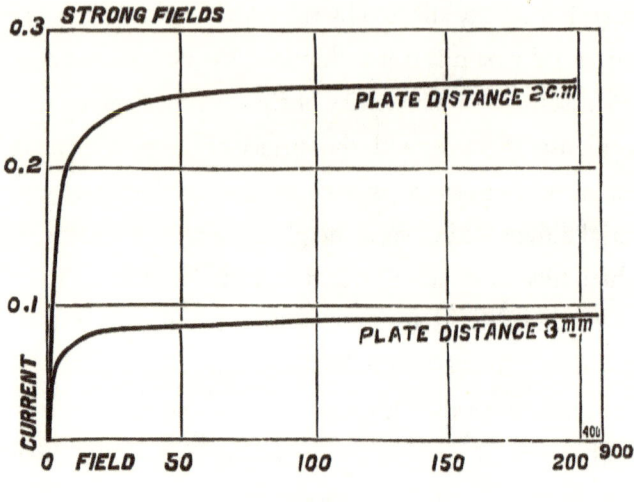

Figure 2.

differences. It is the limiting current that has been adopted as a measure of the radioactivity.

In addition to the difference of potential established between the two plates, there is an electro-motive contact force between them, and these two sources of current combine their effects; this is why the absolute value of the current intensity changes with the sign of the external difference of potential. However, for significant potential differences, the effect of the electro-motive contact force is negligible, and the intensity of the current is then the same, regardless of the direction of the field between the plates.

The study of the conductivity of air and other gases subjected to the action of Becquerel rays has been undertaken by several physicists[10]. A very complete study of the subject has been pub-

---

[10]Becquerel, *Comptes rendus*, Vol. CXXIV, p. 800, 1897. — Kelwin, Beattie and Smolan, *Nature*, Vol. LVI, 1897. — Beattie and Smoluchowskl, *Phil. Mag.*, Vol. XLIII, p. 418.

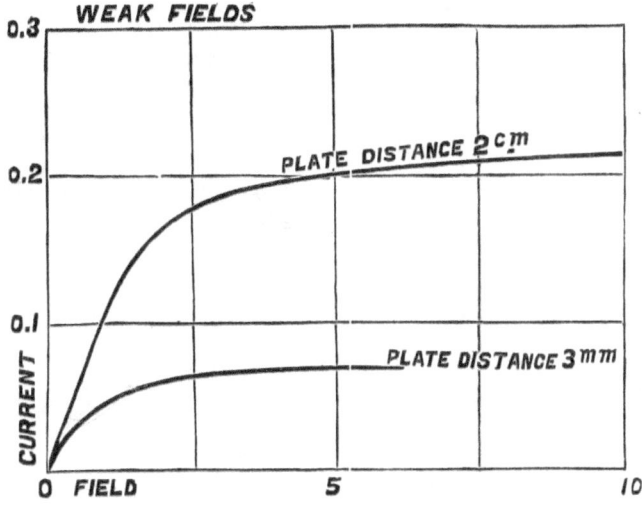

Figure 3.

lished by Mr. Rutherford[11].

The laws of conductivity produced in gases by the Becquerel rays are the same as those found with Röntgen rays. The mechanism of the phenomenon appears to be the same in both cases. The theory of ionisation of gases by the action of the Röntgen or Becquerel rays very well accounts for the facts observed. This theory will not be exposed here. I will only recall the results to which they point: —

Firstly, the number of ions produced per second in the gas is considered to be proportional to the energy of the radiation absorbed by the gas.

Secondly, to obtain the limiting current relating to a given radiation, it is necessary, on the one hand, to have this radiation completely absorbed by the gas, employing a sufficient absorbent mass; on the other hand, it is necessary to use for the production

---

[11]Rutherford, *Phil. Mag.*, January, 1899.

of the current all the ions created, by establishing an electric field strong enough so that the number of ions which recombine becomes a negligible fraction of the total number of ions produced in the same time, which are almost all carried by the current to the electrodes. The higher the electric field required to achieve this result, the higher the amount of ionisation.

According to recent research by Mr. Townsend, the phenomenon is more complex when the gas pressure is low. The current first seems to tend towards a constant limiting value when the potential difference increases; but above a certain potential difference, the current begins again to increase with the field, and with very great rapidity. Mr. Townsend ascribes this increase to a new ionisation produced by the ions themselves when these, under the action of the electric field, take a sufficient speed so that a molecule of gas, encountered by one of these projectiles, becomes broken down into its constituent ions. An intense electric field and a low pressure favour this ionisation by ions already present, and, as soon as this begins to occur, the intensity of the current increases uniformly with the average field between the plates[12]. The limiting current could, therefore, only be obtained under conditions of ionisation, the intensity of which does not exceed a certain value, in such a way that the saturation corresponds to fields for which the shock ionisation of the ions can no longer take place. This condition has occurred in my experiments.

The order of magnitude of the saturation currents obtained with uranium compounds is 10^{-11} amperes for a condenser whose plates have a diameter of 8 c.m. and are 3 c.m. apart. Thorium compounds give rise to currents of the same order of magnitude, and the activity of the uranium and thorium oxides is very similar.

---

[12]Townsend, *Phil. Mag.*, 1901, 6th series, Vol. I, p. 198.

Radioactivity of Uranium and Thorium compounds.

The following are the numbers I obtained with various uranium compounds. I denote the intensity of the current in amperes by the letter i: —

	$i \times 10^{11}$
Metallic uranium (containing a little carbon)	2.3
Black oxide of uranium, U_2O_5	2.6
Green oxide of uranium, U_3O_4	1.8
Hydrated uranic acid	0.6
Uranate of sodium	1.2
Uranate of potassium	1.2
Uranate of ammonium	1.3
Uranium sulphate	0.7
Sulphate of uranium and potassium	0.7
Nitrate of uranium	0.7
Phosphate of copper and uranium	0.9
Oxysulphide of uranium	1.2

The thickness of the layer of the uranium compound used has little effect, provided that the layer is uniform. The following illustrate this point: —

	Thickness of layer M.m.	$i \times 10^{11}$
Uranium oxide	0.5	2.7
Uranium oxide	3.0	3.0
Ammonium uranate	0.5	1.3
Ammonium uranate	3.0	1.4

We can conclude from this, that the absorption of uranium

rays by the substance which emits them is very strong, since the rays coming from the deep layers cannot produce a significant effect.

The numbers I obtained with thorium compounds[13] enable me to state: — Firstly, that the thickness of the layer used has considerable effect, especially with the oxide.

Secondly, that the phenomenon is only regular if a thin active layer is used (e.g., 0.25 m.m.). On the contrary, when we use a thick layer of the substance (6 m.m.), the figures obtained vary between two extreme limits, especially in the case of oxide: —

	Thickness of layer M.m.	$i \times 10^{11}$
Thorium oxide	0.25	2.2
Thorium oxide	0.5	2.5
Thorium oxide	2.5	4.7
Thorium oxide	3.0	5.5 (mean)
Thorium oxide	6.0	5.5 (mean)
Thorium sulphate	0.25	0.8 (mean)

There is in the nature of the phenomenon a cause of irregularities which do not exist in the case of the uranium compounds. The figures obtained for a layer of oxide 6 m.m. thick varied between 3.7 and 7.3.

The experiments that I made on the absorption of uranium and thorium rays showed that thorium rays are more penetrating than uranium rays, and that the rays emitted by the thorium oxide in a thick layer are more penetrating than those it emits in a thin layer. The following figures give the fraction of the radiation transmitted by a sheet of aluminium with a thickness of 0.01 m.m.

---

[13] Mme Curie, *Comptes rendus*, April 1898.

With uranium compounds, the absorption is the same regardless of the compound used, which suggests that the rays emitted by the various compounds are of the same nature.

Radioactive substance	Fraction of radiation transmitted by sheet
Uranium	0.18
Uranium oxide, U_2O_5	0.20
Uranate of ammonium	0.20
Phosphate of uranium and copper	0.21
Thorium oxide of thickness 0.25 m.m.	0.38
Thorium oxide of thickness 0.5 m.m.	0.47
Thorium oxide of thickness 3.0 m.m.	0.70
Thorium oxide of thickness 0.60 m.m.	0.70
Thorium sulphate of thickness 0.25 m.m.	0.38

The characteristics of the thorium radiation have been the subject of very comprehensive publications. Mr. Owens[14] has demonstrated that the uniform current is only obtained after a fairly long time in an enclosed apparatus, and that the intensity of the current is greatly reduced under the action of a current of air (which does not happen for uranium compounds). Mr. Rutherford has made similar experiments, and has interpreted them by asserting that thorium and its compounds emit not only the Becquerel rays, but also an *emanation*, composed of extremely minute particles, which remain radioactive for some time after their emission, and are capable of being swept along by a current of air[15].

The characteristics of the thorium radiation, which are relative to the influence of the thickness of the layer employed and to the action of air currents, have a close connection with the pheno-

---

[14]Owens, *Phil. Mag.*, October 1899.
[15]Rutherford, *Phil. Mag.*, January 1900.

menon of the *radioactivity induced, and of its gradual diffusion*. This phenomenon was observed for the first time with radium, and will be described later.

The radioactivity of thorium and uranium compounds appears as an *atomic property*. M. Becquerel had previously observed that all uranium compounds are active, and concluded that their activity was due to the presence of the element uranium; he also demonstrated that uranium was more active than its salts[16]. From this point of view I have studied the compounds of thorium and uranium, and I have made a large number of measurements of their activity under various conditions. It follows from all of these measurement that the radioactivity of these substances is indeed an atomic property. It seems here to be linked with the presence of atoms of the two elements considered, and it is not destroyed either by changes of the physical state or by chemical transformations. The chemical combinations and mixtures containing uranium or thorium are more active as they contain a higher proportion of these metals, all inactive material acting as both inert bodies and absorbing material.

Is Atomic Radioactivity a general Phenomenon?

As mentioned above, I made experiments to look for radioactive substances other than compounds of uranium and thorium. I started this research with the idea that it was very unlikely that radioactivity, considered to be an atomic property, should belong to a certain kind of matter to the exclusion of all others. The measurements I have made allow me to say that, for the chemical elements currently considered as such, including the rarest and the most hypothetical, the compounds I investigated have always

---

[16]Becquerel, *Comptes rendus*, Vol. CXXII, 1896.

been at least 100 times less active in my device than metallic uranium. In the case of widespread elements, I have studied several compounds; in the case of rare bodies, I studied the compounds that I was able to obtain.

The following is a list of the substances that were part of my study either as the element or in combination: —

1. All the metals or metalloids that are easily found, and some, more rare, pure products, from the collection of M. Etard, at the Ecole de Physique et de Chimie Industrielles de la Ville de Paris.

2. The following rare bodies: — gallium, germanium, neodymium, praseodymium, niobium, scandium, gadolinium, erbium, samarium, and rubidium (specimens lent by M. Demarçay), yttrium, ytterbium (lent by M. Urbain[17]).

3. A large number of rocks and minerals.

Within the limits of sensitivity of my apparatus, I have found no simple substance other than uranium and thorium, possessing atomic radioactivity. However, a few words should be said here concerning phosphorus. White moist phosphorus, placed between the plates of the condenser, causes the air between the plates to conduct[18]. However, I do not consider this body to be radioactive like uranium and thorium. Phosphorus, in fact, under these conditions, oxidises and emits light rays, whilst uranium and thorium compounds are radioactive without experiencing any appreciable chemical modification detectable by known

---

[17]I am very grateful to the scholars mentioned above, to whom I owe samples that were used for my study. I thank also M. Moissan who kindly donated some metallic uranium for this study.

[18]Elster and Geitel, *Wied. Ann.*, 1890

means. In addition, phosphorus is neither active in the red variety state, nor in the combination state.

In a recent work, M. Bloch has demonstrated that phosphorus, by oxidising in air, gives rise to very little mobile ions, which make the air conductive and cause the condensation of water vapour[19].

Some recent work would lead to the conclusion that radioactivity belongs to all substances to an extremely low degree[20]. The identification of these very weak phenomena with the phenomena of atomic radioactivity cannot yet be considered established.

Uranium and thorium are the two elements with the highest atomic weights (240 and 232); they are frequently found in the same minerals.

Radioactive Minerals.

I have examined many minerals in my apparatus[21]; some of them have been shown to be radioactive, e.g., pitchblende, thorite, orangite, fergusonite, cleveite, chalcolite, autunite, monazite, &c. The following is a table giving in amperes the intensity, i, of the current obtained with metallic uranium and with various minerals: —

	$i \times 10^{11}$
Uranium	2.3
Pitchblende from Johanngeorgenstadt	8.3
Pitchblende from Joachimsthal	7.0
Pitchblende from Pzibran	6.5
Pitchblende from Cornwallis	1.6
Cleveite	1.4

[19]Bloch, *Société de Physique*, 6 February 1908.
[20]Mac Lennan and Burton, *Phil. Mag.*, June 1903. – Strutt, *Phil. Mag.*, June 1903. – Lester Cooke, *Phil. Mag.*, October 1903.
[21]Several samples of minerals from the Museum's collection have been kindly made available to me by M. Lacroix.

Chalcolite	5.2
Autunite	2.7
Various thorites	$\begin{cases} 0.3 \\ 0.7 \\ 1.3 \\ 1.4 \end{cases}$
Orangite	2.0
Monazite	0.5
Xenotime	0.03
AEschynite	0.7
Fergusonite (two samples)	$\begin{cases} 0.4 \\ 0.1 \end{cases}$
Samarskite	1.1
Niobite (two samples)	$\begin{cases} 0.1 \\ 0.3 \end{cases}$
Tantalite	0.02
Carnotite[22]	6.2

The current obtained with orangite (thorium oxide ore) varied greatly with the thickness of the layer used. By increasing this thickness from 0.25 m.m. to 6 m.m. the current increased from 1.8 to 2.3.

All minerals that are radioactive contain uranium or thorium: their activity is therefore not surprising, but the intensity of the phenomenon in certain cases is unexpected. Thus there are pitch-blendes (uranium oxide ores) which are four times as active as metallic uranium. Chalcolite (double phosphate of copper and uranium) is twice as active as uranium. Autunite (phosphate of uranium and calcium) is as active as uranium. These facts were at odds with the previous conclusions, according to which no mineral should have been more active than uranium or thorium.

---

[22]Carnotite is a mineral of uranate vanadate recently discovered by Friedel and Cumenge.

To clarify this point, I prepared artificial chalcolite by the process of Debray, starting from pure products. The process consists in mixing a solution of uranium nitrate with a solution of copper phosphate in phosphoric acid and heating to 50° or 60°. After some time, crystals of chalcolite form in the liquid[23]. The chalcolite thus obtained has a completely normal activity, given by its composition; it is two and a half times less active than uranium.

It therefore appeared very likely that if pitchblende, chalcolite and autunite have such strong activity it is because these substances contain a small quantity of a strongly radioactive material, different from uranium, thorium and the simple bodies currently known. I thought that if this were indeed the case, I could hope to extract this substance from the ore by the ordinary methods of chemical analysis.

---

[23]Debray, *Ann. de Chim. et de Phys.*, 3rd series, Vol. LXI, p. 415.

Chapter II. The New Radioactive Substances.

Methods of research.

The results of the study of radioactive minerals, set out in the preceding chapter, prompted M. Curie and myself to seek to extract a new radioactive substance from pitchblende. Our research method could only be based on radioactivity, as we knew of no other property of the hypothetical substance. The following is the method pursued for a research based on radioactivity: — The radioactivity of a compound is determined, and a chemical separation is carried out on this product; the radioactivity of all the products obtained is determined, and one discerns whether the radioactive substance has remained entirely with one of them, or whether it has been distributed among them and in what proportion. In this way, an indication is obtained, which may to a certain extent be compared to that which spectral analysis could provide. In order to obtain comparable numbers, the activity must be measured for solid, well-dried substances.

Polonium, Radium, Actinium.

The analysis of pitchblende with the help of the method just explained, led us to establish the existence, in this mineral, of two strongly radioactive substances, chemically different: — *Polonium*, discovered by ourselves, and *radium*, which we discovered in collaboration with M. Bémont[1].

Polonium from the analytical point of view, is similar to bismuth, and separates out with the latter. One obtains bismuth increasingly rich in polonium by one of the following methods of fractionating: —

1. Sublimation of sulfides *in vacuo*; active sulfide is much more volatile than bismuth sulfide.

2. Precipitation of nitrogenous solutions by water; the precipitate of the basic nitrate is much more active than the salt which remains in solution.

3. Precipitation by sulphuretted hydrogen of an extremely acidic hydrochloric solution; the precipitated sulfides are considerably more active than the salt which remains in solution.

Radium is a substance that accompanies the barium removed from pitchblende; it resembles barium in its reactions, and is separated from it by difference of solubility of the chlorides in water, in dilute alcohol, or in water with hydrochloric acid. We effect the separation of the chlorides of barium and radium by subjecting the mixture to fractional crystallisation, radium chloride being less soluble than that of barium.

A third highly radioactive substance has been identified in pitchblende by M. Debierne, who gave it the name of *actinium*[2]. Actinium accompanies certain members of the iron group contained in pitchblende; it seems particular close to thorium, from

---

[1] P. Curie and Mme Curie, *Comptes rendus*, July 1898. — P. Curie, Mme Curie and G. Bémont, *Comptes rendus*, December 1898.
[2] Debierne, *Comptes rendus*, October 1899 and April 1900.

which it has not yet been found possible to separate it. The extraction of actinium from pitchblende is a very difficult operation, the separations being generally incomplete.

All three new radioactive substances are found in pitchblende in absolutely infinitesimal amounts. In order to obtain them in a more concentrated state, we had to undertake the processing of several tons of uranium ore residue. The big treatment was carried out in the factory; and this was followed by processes of purification and concentration. We thus succeeded in extracting from thousands of kilogrms. of raw material a few decigrammes of products which were exceedingly active compared to the ore from which they were obtained. It is quite obvious that this process is long, arduous, and expensive.[3]

Other new radioactive bodies have also been reported since the termination of our work. M. Giesel, on the one hand, and MM. Hoffmann and Strauss on the other, have announced the probable existence of a radioactive substance similar to lead in its chemical properties. At present only a few samples of this substance have been obtained[4].

---

[3]We are much obliged to all who came to our aid in this work. We sincerely thank MM. Mascart and Michel Lévy for their benevolent support. We thank the benevolent intervention of Professor Suess, the Austrian government that has graciously provided us with the first tonne of residue treated (from state plant, at Joachimsthal, in Bohemia). The Paris Academy of Sciences, the National Industry Incentive Society, an anonymous donor, thanks to whom we have been given the opportunity to process a certain amount of product. Our friend, Mr. Debierne, organized the treatment of ore, which was carried out in the plant of the Central Company of Chemical Products. This Company has consented to carry out the processing without looking for a profit. To all, we address our sincere thanks. More recently, the Institut de France has provided us with a sum of 20,000fr for the extraction of radioactive materials. Thanks to this sum, we were able to start processing 5 tons of ore.
[4]Giesel, *Ber. deutsch. chem. Gesell.*, Vol. XXXIV, 1901, p. 3775. — Hoffmann and Strauss, *Ber. deutsch. chem. Gesell.*, Vol. XXXIII, 1900, p. 3126.

Of all the new radioactive substances, radium has so far been
the only one that has been isolated in the form of pure salt.

Spectrum of Radium.

It was of primary importance to check, by all possible means, the
hypothesis, underlying this work, of new radioactive elements.
The spectral analysis was used, in the case of radium, to confirm
this hypothesis.

M. Demarçay has been kind enough to undertake the examina-
tion of the new radioactive substances by the rigorous procedures
that he employs in the study of photographic spark spectra.

The assistance of such a competent scientist has been a great
blessing to us, and we are deeply grateful to him for having consented to take up this work. The results of the spectral analysis
brought conviction to us, while we were still in doubt about the
interpretation of the results of our research[5].

The first samples of fairly active barium chloride containing
radium, examined by M. Demarçay, showed him, together with
the barium lines, a new line of considerable intensity and of wave-
length $\lambda = 381.47 \ \mu\mu$ in the ultra-violet. With more active products, prepared subsequently, Demarçay saw the line 381.47 $\mu\mu$
becoming stronger; at the same time other new lines appeared,
and the intensity of the new lines became comparable with that
of the barium lines. A new concentration provided a product
for which the new spectrum dominated, and the three strongest
barium lines, alone visible, merely indicated the presence of this

---

[5]Just recently we had the pain of witnessing the departure of a scholar so
distinguished, as he continued his fine research on rare earths and on spec-
troscopy, whose perfection and precision in his methods we cannot admire
too much. We keep a fond memory of the perfect kindness with which he
had consented to take part in our work.

metal as an impurity. This product may be considered as nearly pure radium chloride. Finally, by further purification, I was able to obtain an extremely pure chloride, in the spectrum of which the two dominant barium lines were barely visible.

The following is a list, according to Demarçay[6], of the main radium lines for the portion of the spectrum between λ = 500.0 and λ = 350.0 $\mu\mu$. The intensity of each line is indicated by a number, the strongest line being marked 16: —

λ.	Intensity	λ.	Intensity
482.63	10	460.03	3
472.69	5	453.35	9
469.98	3	443.61	8
469.21	7	434.06	12
468.30	14	381.47	16
464.19	4	364.96	12

All the lines are sharp and narrow, the three lines 381.47, 468.30, 434.06 are strong, and equal the most intense of those currently known. Two strong misty bands are also visible in the spectrum. The first, symmetrical, ranges from 463.10 to 462.19, with a maximum at 462.75. The second, stronger, fades towards the ultra-violet; it starts, sharply defined, at 446.37, and passes through a maximum at 445.52; the region of the maximum extends as far as 445.34, then a nebulous band, gradually fading, extends until around 439.

In the least refrangible part, not photographed in the spark spectrum, the only significant line is 566.5 (approx.), which is much fainter, however, than 482.63.

The general appearance of the spectrum is that of the alkaline earth metals; these metals are known to have strong line spectra

---

[6]Demarçay, *Comptes rendus*, December 1898, November 1899 and July 1900.

with some nebulous bands.

According to Demarçay, radium can be among the bodies possessing the most sensitive spectral reaction. I have, moreover, been able to conclude from my work of concentration, that, in the first sample examined, which clearly showed the line 3814.7, the proportion of radium must have been very small (perhaps about 0.02 per cent). Nevertheless, it takes an activity fifty times as great as that of metallic uranium to clearly see the main radium line in the spectra photographed. With a sensitive electrometer, we can detect the radioactivity of a substance when it is only 1/100 of that of metallic uranium. It is clear that, in order to detect the presence of radium, the property of radioactivity is several thousand times more sensitive than the spectral reaction.

The very active bismuth containing polonium and the very active thorium containing actinium, examined by Demarçay, have so far given only the lines of bismuth and thorium, respectively.

In a recent publication, M. Giesel[7], who has been involved in the preparation of radium, states that radium bromide gives rise to a carmine coloring of the flame. The radium flame spectrum contains two beautiful red bands, one line in the blue-green, and two faint lines in the violet.

Extraction of the New Radioactive Substances.

The first stage of the operation consists in extracting from the uranium ores the barium with radium, the bismuth with polonium and the rare earths containing actinium. These three first products having been obtained, the next step is in each case to endeavor to isolate the new radioactive substance. This second part of the treatment is done by a fractionation method. We

---

[7]Giesel, *Phys. Zeitschrift*, 15 September 1902.

know that it is difficult to find a very perfect means of separation between very similar elements; methods of fractionation are therefore ideal. Besides this, when a mere trace of one element is mixed with another element, one cannot apply a perfect separation method to the mixture, even assuming that such a method is known; in fact, we risk losing the trace of material that could have been separated in the operation.

I was especially involved in the work aimed at isolating radium and polonium. After a few years of work, I have so far only succeeded in obtaining the former.

As pitchblende is an expensive ore, we have given up processing large quantities. In Europe the ore is mined at the Joachimsthal mine, in Bohemia. The crushed ore is roasted with sodium carbonate, and the resulting material washed, first with warm water and then with dilute sulphuric acid. The solution contains uranium, which gives pitchblende its value. The insoluble residue is discarded. This residue contains radioactive substances; its activity is four and a-half times greater than that of metallic uranium. The Austrian government, to whom the mine belongs, has kindly given us a ton of this residue for our research, and has authorised the mine to supply us with several more tons of the material.

It was not very easy to apply the first treatment of the residue at the factory by following the same procedure as in the laboratory. M. Debierne was kind enough to investigate this question, and organised the treatment in the factory. The most important point of his method which he indicated consists in the conversion of the sulphates into carbonate by boiling the material with a concentrated solution of sodium carbonate. This method avoids the necessity of fusing with sodium carbonate.

The residue mainly contains the sulphates of lead and calcium, silica, alumina, and iron oxide. In addition, nearly all the metals are found in greater or smaller amount (copper, bismuth, zinc, cobalt, manganese, nickel, vanadium, antimony, thallium, rare earths, niobium, tantalum, arsenic, barium, &c.). Radium is found in this mixture in the sulphate state, and constitutes its least soluble sulphate. In order to dissolve it, it is necessary to remove the as much sulfuric acid as possible. To do this, we start first by treating the residue with a boiling concentrated solution of ordinary soda. The sulfuric acid combined with the lead, aluminium, and calcium passes, for the most part, into solution as a sodium sulphate, which is removed by repeatedly washing with water. The alkaline solution removes at the same time lead, silicon, and aluminium. The insoluble portion washed in water is attacked by ordinary hydrochloric acid. This operation completely disaggregates the material, and dissolves a large part of it. Polonium and actinium can be removed from this solution; the former is precipitated by sulphuretted hydrogen, the latter is found in the hydrates precipitated by ammonia in the solution separated from the sulphides and oxidised. As of radium, it remains in the insoluble portion. This portion is washed with water, and then treated with a boiling concentrated solution of sodium carbonate. This operation completes the transformation of the sulphates of barium and radium into carbonates. The material in then washed very thoroughly with water, and then treated with dilute hydrochloric acid, free from sulphuric acid. The solution contains radium as well as polonium and actinium. It is filtered and precipitated by sulphuric acid. This produces crude sulphates of barium containing radium, calcium, lead, iron, and also a trace of actinium. The solution still contains a little actinium and polonium, which can be removed as in the case of the first hydrochloric acid solution.

We extract from one ton of residue 10 to 20 kilogrms. of crude sulphates, whose activity is from thirty to sixty times as great as that of metallic uranium. They must now be purified. To do this, they are boiled with sodium carbonate and transformed into chlorides. The solution is treated with sulphuretted hydrogen, which gives a small amount of active sulphides containing polonium. The solution is filtered, oxidised by the action of chlorine, and precipitated by pure ammonia. The precipitated oxides and hydrates are very active, and the activity is due to actinium. The filtered solution is precipitated by sodium carbonate. The precipitated alkaline earth carbonates are washed and transformed into chlorides. These chlorides are evaporated to dryness, and washed with pure concentrated hydrochloric acid. The calcium chloride dissolves almost entirely, whilst the chloride of barium and radium remains insoluble. Thus, from one ton of the original raw material about 8 kilogrms. of barium and radium chloride are obtained, the activity of which is about sixty times that of metallic uranium. The chloride is now ready for fractionation.

Polonium.

As mentioned above, by passing sulphuretted hydrogen through the various hydrochloric acid solutions obtained during the treatment, active sulphides are precipitated, the activity of which is due to polonium. These sulfides contain mainly bismuth, a little copper and lead; the latter metal is not found in high proportion there, because it has been largely removed by the soda solution, and because its chloride is not very soluble. Antimony and arsenic are found in oxides only in small quantities, their oxides having been dissolved by soda. In order to obtain immediately very active sulfides, the following process was employed:

— The very acidic hydrochloric solutions were precipitated

by sulphuretted hydrogen; the sulfides thus precipitated are very active, and are used for the preparation of polonium; there remain in the solution substances not completely precipitated in presence of excess of hydrochloric acid (bismuth, lead, antimony). To complete the precipitation, the solution is diluted with water, and treated again with sulphuretted hydrogen, which gives a second precipitate of sulphides, much less active than the first, and which have generally been rejected. For further purification of the sulfides, they are washed with ammonium sulfide, which removes the remaining traces of antimony and arsenic. Then they are washed with water containing ammonium nitrate, and treated with dilute nitric acid. Dissolution is never complete; there is always an important insoluble residue, which can be treated again if deemed useful. The solution is reduced to a small volume and precipitated either by ammonia or by excess of water. In both cases lead and copper remain in solution; in the second case, a little bismuth, barely active, remains also in solution.

The precipitate of oxides or basic nitrates is subjected to fractionation in the following manner: — The precipitate is dissolved in nitric acid, and water is added to the solution until a sufficient quantity of precipitate is formed; it is necessary to take into account that sometimes the precipitate does not appear at once. The precipitate is separated from the supernatant liquid, and re-dissolved in nitric acid, after which both the liquids thus obtained are re-precipitated with water, and treated as before. The different fractions are combined according to their activity, and concentration is carried out as far as possible. We thus obtain a very small quantity of matter whose activity is enormous, but which, nevertheless, has not yet shown bismuth lines in the spectroscope.

Unfortunately, by this means, it is unlikely to result in the isolation of polonium. The method of fractionation just described presents great difficulties, and the same is true with other wet processes of fractionation. Whatever process is used, compounds are readily formed which are absolutely insoluble in dilute or concentrated acids. These compounds can only be re-dissolved by reducing them to the metallic state, e.g., by fusion with potassium cyanide. Given the considerable number of operations necessary, this circumstance constitutes an enormous difficulty for the progress of the fractionation. This drawback is all the more serious since polonium is a substance which, once extracted from pitchblende, decreases in activity. This decrease in activity is also slow: a sample of bismuth nitrate containing polonium only lost half its activity in eleven months.

No similar difficulty arises with radium. The radioactivity remains throughout an accurate gauge of the concentration; the concentration itself presents no difficulty, and the progress of the work from the beginning can be constantly checked by spectral analysis.

When the phenomena of induced radioactivity, which will be discussed later on, were known, it seemed natural to assume that polonium, which only shows the bismuth lines and whose activity decreases over time, was not a new element, but bismuth made active by the vicinity of radium in the pitchblende. I am not convinced that this view is correct. In the course of my extended work on polonium, I have seen chemical effects that I have never observed with ordinary bismuth or with radium-activated bismuth. These chemical effects are, in the first place, the extremely easy formation of insoluble compounds, of which I have spoken above (especially basic nitrates), and, in the second place, the colour and appearance of the precipitates obtained by adding water to the nitric acid solution of bismuth containing polonium.

These precipitates are sometimes white, but more generally of a more or less bright yellow, verging on dark red.

The absence of lines other than those of bismuth does not conclusively prove that the substance only contains bismuth, because there are bodies whose spectral reaction is scarcely visible.

It would be necessary to prepare a small amount of bismuth containing polonium in a state of concentration as high as possible, and to study it chemically, first by determining the atomic weight of the metal. It has not yet been possible to carry out this research on account of the difficulties of a chemical nature already mentioned.

If polonium were proved to be a new element, it would not be less true that this element cannot exist indefinitely in the highly radioactive state, at least when extracted from the ore. We can then consider the question in two different ways: — First, whether the activity of polonium is entirely induced by the proximity of substances themselves radioactive, in which case polonium would possess the faculty of acquiring atomic activity in a durable way, a faculty which does not seem to belong to any substance; second, whether the activity of polonium is an inherent property, which is spontaneously destroyed under certain conditions, and can persist under certain other conditions which are found in the ore. The phenomenon of atomic activity induced by contact is still so poorly understood, that there is a lack of a basis for forming a coherent opinion on the matter.

(Note. — A work has recently appeared on polonium by M. Marckwald[8]. He plunges a small rod of pure bismuth into a hydrochloric acid solution of the bismuth extracted from the pitchblende residue. After some time the rod becomes coated

---

[8]*Berichte d. deutsch. chem. Gesell.*, June 1902 and December 1902.

with a very active deposit, and the solution contains only inactive bismuth. M. Marckwald also obtains a very active deposit by adding tin chloride to a hydrochloric acid solution of radioactive bismuth. From this he concludes that the active element is analogous to tellurium, and gives it the name of radiotellurium. This active substance of M. Marckwald appears to be identical to polonium, from its behaviour, and from the highly absorbable rays it emits. The choice of a new name for this material is futile in the current state of the question).

Preparation of the Pure Chloride of Radium.

The process that I adopted to extract pure radium chloride from barium chloride containing radium consists in subjecting the mixture of chlorides to a fractional crystallisation in pure water first, then in water added with pure hydrochloric acid. The difference in solubility of the two chlorides is thus made use of, that of radium being less soluble than that of barium.

At the beginning of the fractionation, pure distilled water is used. The chloride is dissolved, and the solution is brought to saturation at boiling temperature, then allowed to crystallise by cooling in a covered capsule. Beautiful crystals form at the bottom, and the supernatant, saturated solution can be easily decanted. If part of this solution is evaporated to dryness, it is found that the chloride obtained is approximately five times less active than that which has crystallised. The chloride is thus divided into two portions, *A* and *B* — portion *A* being more active than portion *B*. The same operation is repeated on each of the chlorides *A* and *B*, and in each case two new portions are obtained. When the crystallisation is complete, the least active fraction of chloride *A* and the most active fraction of chloride *B* are combined together, these two materials having approximately the same activity. We then find ourselves having three portions which we submit again to the same treatment.

The number of portions is not allowed to increase indefinitely. As this number increases, the activity of the most soluble portion decreases. When this portion has nothing more than insignificant activity, it is eliminated from the fractionation. When the desired number of fractions has been obtained, fractionation of the least soluble portion is stopped (the richest in radium), and it is eliminated from the fractionation.

We operate with a constant number of portions. After each series of operations, the saturated solution from one fraction is poured onto the crystals arising from the following fraction; but if after one of the series, the most soluble fraction has been eliminated, then, after the following series, a new fraction is made from the most soluble portion, and the crystals of the most active portion are removed. By the alternative succession of these two processes, a very regular fractionation mechanism is obtained, in which the number of fractions and the activity of each remains constant, each being about five times more active than the next, and in which, on the one hand, an almost inactive product is removed, whilst, on the other, is obtained a chloride rich in radium. The amount of material contained in these fractions gradually decreases, becoming less as the activity increases.

At first six fractions were used, and the activity of the chloride obtained at the end was only 0.1 that of uranium.

When most of the inactive matter has been removed, and the fractions have become small, one fraction is removed from the one end, and another is added to the other end consisting of the active chloride previously removed. We will therefore now collect a chloride richer in radium than previously. This system continues to be applied until the top crystals obtained are pure radium chloride. If the fractionation has been done in a very thorough way, there are hardly any traces of the intermediate products.

When the fractionation is in an advanced stage and the quantity of material in each fraction is small, the separation by crystallisation is less effective, the cooling being too rapid and the volume of the solution to be decanted too small. It is therefore advantageous to add water containing a known quantity of hydrochloric acid; this quantity may be increased as the fractionation progresses.

The advantage of this addition consists in increasing the quantity of the solution, the solubility of chlorides being less in water acidified with hydrochloric acid than in pure water. By using water with a lot of acid, you have excellent separations, and you can operate with only three or four fractions. It is best to use this process as soon as the amount of material has become small enough that it can be done without disadvantages.

The crystals, which form in a very acid solution, have the shape of very elongated needles, those of barium chloride having exactly the same appearance as those of radium chloride. Both are birefringent. The crystals of barium chloride containing radium are colourless, but when the proportion of radium becomes sufficient, they have a yellow colouration after some hours, verging on orange, and sometimes a beautiful pink. This colour disappears in solution. The crystals of pure radium chloride do not colour, or at least not as quickly, so that the colouration appears to be due to the simultaneous presence of barium and radium. The maximum colouration is obtained for a certain concentration of radium, and it is possible, based on this property, to monitor the progress of the fractionation. As long as the most active portion becomes colored, it contains a significant amount of barium; when it is no longer colored, and the following portions are colored, it means that the first is substantially pure radium chloride.

I have sometimes noticed the formation of a deposit composed of crystals, part of which remains colourless, whilst the other becomes coloured, and it seems possible that the colourless crystals might be sorted out.

The fractional precipitation of an aqueous solution of barium chloride by alcohol also leads to the isolation of radium chloride, which is the first to precipitate. This method, which I used at the beginning, was then abandoned for the one that has just been exposed and which offers more regularity. However, I have still sometimes used alcohol precipitation to purify radium chloride which contains traces of barium chloride. The latter remains in the slightly aqueous alcoholic solution, and can thus be removed.

M. Giesel, who, as soon as our first research was published, took care of the preparation of radioactive bodies, recommends the separation of barium and radium by fractional crystallisation in water from a mixture of bromides. I can testify that this method is indeed very advantageous, especially in the first stages of the fractionation.

Whatever fractionation process is used, it is useful to control it by activity measurements.

It is necessary to note that a radium compound which was dissolved, and which then was brought back to the solid state, either by precipitation, or by crystallization, has at the beginning an activity which is smaller as it stays longer in dissolution. The activity then increases for several months to reach a certain limit, always the same. The final activity is five to six times higher than the initial activity. These variations, to which I will return later, must be taken into account for the measurement of activity. Although the final activity is better defined, it is more practical, during chemical treatment, to measure the initial activity of the solid product.

The activity of highly radioactive substances is of a completely different order of magnitude than that of the ore from which they come (it is 10^6 times greater). When we measure this radioactivity by the method that was exposed at the beginning of this work (apparatus in Fig. 1), we cannot increase, beyond a certain limit, the charge that we put in the quartz plate. This charge, in our experiments, reaches a maximum for 4000 g, corresponding to an amount of released electricity equal to 25 electrostatic units. We can measure activities that vary, in the ratio of 1 to 4000, always using the same area for the active substance. To extend the limits of the measurements, we vary this area in a known ratio. The active substance then occupies on the tray B a central circular area of known radius. The activity not being, under these conditions, exactly proportional to the surface, we determine experimentally coefficients which allow to compare the activities with uneven active surface.

When this resource itself is depleted, we are forced to resort to the use of absorbent screens and other equivalent processes which I will not insist on here. All these procedures, more or less imperfect, are however sufficient to guide research.

We also measured the current flowing through the capacitor when it is switched on with a battery of small accumulators and a sensitive galvanometer. The need to frequently check the sensitivity of galvanometer prevented us from using this method for routine measurements.

Determination of the Atomic Weight of Radium.[9]

In the course of my work I have repeatedly determined the atomic weight of the metal contained in specimens of barium chloride containing radium. With each newly obtained product I carried the concentration as far as possible, so as to obtain from 0.1 grm. to 0.5 grm. of material containing almost all of the activity of the mixture. From this small quantity of matter I precipitated with alcohol or with hydrochloric acid some milligrms. of chloride for spectral analysis. Thanks to his excellent method, Demarçay needed only this small quantity of material to obtain the photograph of the spark spectrum. I made an atomic weight determination with the product I had left.

I used the classic method which consists in dosing, in the state of silver chloride, the chlorine contained in a known weight of the anhydrous chloride. As a control experiment, I determined the atomic weight of barium by the same method, under the same conditions, and with the same amount of material, first 0.5 grm. and then 0.1 grm. The numbers found were always between 137 and 138. I saw that this method gives satisfactory results, even with a such a small amount of material.

The first two determinations were made with chlorides, of which one was 230 times and the other 600 times as active as uranium. These two experiments gave the same measurement accuracy as the experiment with the pure barium chloride. Therefore the only hope to find a difference relied on using a much more active product. The following experiment was made with a chloride, whose activity was about 3500 times times larger than that of uranium; and this experiment made it possible, for the

---

[9]Mme Curie, *Comptes rendus*, 13 November 1899, August 1900 and 21 July 1902.

first time, to observe a small but distinct difference; I found, as the mean atomic weight of the metal contained in this chloride, the number 140, which indicated that the atomic weight of radium must be higher than that of barium. By using more and more active products, and obtaining spectra of radium of increasing intensity, I noted that the numbers obtained were also increasing, as is seen in the following table (*A* indicates the activity of chloride, that uranium being taken as a unit; *M* the atomic weight found).

A.	M.	
3500	140	Spectrum of radium faint
4700	141	
7500	145.8	Spectrum of radium strong, but that of barium predominating
Order of magnitude 10^6	173.8	The two spectra of almost equal intensity
Order of magnitude 10^6	225	Only a trace of barium present

The numbers of column A should only be taken as a rough indication. It is difficult to assess the activity of highly radioactive bodies, for various reasons which will be discussed below.

At the termination of the processes described above, I obtained, in March, 1902, 0.12 grm. of radium chloride, of which Demarçay was kind enough to make the spectral analysis. This radium chloride, according to the opinion of Demarçay, was substantially pure; however, its spectrum still showed the three principal barium lines with considerable intensity. I made four successive estimations of the chloride, the results of which are as follows: —

	Anhydrous radium chloride.	Silver chloride.	M.
I.	0.1150	0.1130	220.7
II.	0.1140	0.1119	223.0
III.	0.11135	0.1086	222.8
IV.	0.10925	0.10645	223.1

I then re-purified this chloride, and obtained an even purer substance, in the spectrum of which the two strongest barium lines were very faint. Given the sensitivity of the spectral reaction of barium, Demarçay estimated that this purified chloride contained only "minimal traces of barium, incapable of influencing the atomic weight to an appreciable extent". I made three determinations with this perfectly pure radium chloride. The results were as follows: —

	Anhydrous radium chloride.	Silver chloride.	M.
I.	0.09192	0.08890	225.3
II.	0.08936	0.08627	225.8
III.	0.08839	0.08589	224.0

The numbers give an average of 225. They were calculated in the same way as the preceding ones by considering radium as a bivalent element, the chloride having the formula $RaCl_2$, and by adopting for silver and chlorine the values $Ag = 107.8$, $Cl = 35.4$.

It follows that the atomic weight of radium is $Ra = 225$. I consider this number to be exact to the nearest unit.

The weighings were made with a Curie aperiodic balance, perfectly regulated, accurate to the twentieth of a milligrm. This scale, with direct reading, allows very rapid weighing, which is an essential condition in the case of the anhydrous chlorides of radium and barium, which gradually absorb moisture, despite the presence of drying bodies in the balance. The materials to be weighed were placed in a platinum crucible; this crucible had been in use for a long time, and I verified that its weight did not vary by the tenth part of a milligrm. during the course of one operation.

The hydrated chloride obtained by crystallisation was placed into the crucible and heated in the oven to be transformed into anhydrous chloride. Experience has shown that when the chloride has been kept for several hours at 100° its weight becomes constant, and does not change even if the temperature is raised to 200° and held there for a few hours. The anhydrous chloride thus obtained constitutes, therefore, a perfectly defined body.

The following is a series of determinations relating to this subject. The chloride (100 m.g.) is dried in the oven at 55°, and placed in a desiccator over anhydrous phosphoric acid; it then loses weight very slowly, which proves that it still contains moisture; in the course of twelve hours the loss was 3 m.g. The chloride is transferred to the oven, and the temperature raised to 100°. During this process, the chloride lost 6.3 m.g. in weight. After being left three hours fifteen minutes in the oven, it lost 2.5 m.g. more. The temperature was maintained for forty-five minutes between 100° and 120°, which caused a loss of weight of 0.1 m.g. Then left for thirty minutes at 125°, the chloride does not lose anything. Maintained then for thirty minutes at 150°, it lost 0.1 m.g. Finally, after being heated for four hours at 200°, it lost 0.15 m.g. During these operations the crucible varied from 0.05 m.g.

After each determination of the atomic weight, the radium was brought back to the chloride state in the following manner: — To the solution containing the weighed radium nitrate and excess of silver nitrate was added pure hydrochloric acid; the silver chloride was filtered off; the solution was evaporated to dryness several times with excess of pure hydrochloric acid. In this way the nitric acid is entirely removed.

The precipitated silver chloride was always radioactive and phosphorescent. In determining the amount of silver contained in it, I made sure it had not resulted in any reasonable amount of radium being carried down with it out of the solution. The method I pursued was to reduce the silver chloride precipitated in the crucible by hydrogen generated from dilute hydrochloric acid and zinc; after washing, the crucible was weighed with the metallic silver contained therein.

I also found, in one experiment, that the weight of the regenerated radium chloride was the same as before the beginning of the operation. In other experiments, I did not wait, to start a new operation, that all the washing water was evaporated.

These verifications do not have the same precision as direct experiments; however, they ensured that no significant mistakes were made.

Based on its chemical properties, radium is an element of the group of alkaline earths, being the member next above barium.

From its atomic weight, radium is also placed in Mendeleeff's table after barium with the alkaline earth metals, in the row which already contains uranium and thorium.

Characteristics of the Radium Salts.

The salts of radium, chloride, nitrate, carbonate, and sulphate, resemble those of barium, when they have been prepared in the solid state, but they gradually become coloured.

All the radium salts are bright in the dark.

By their chemical properties, the salts of radium are absolutely analogous to the corresponding salts of barium. However, radium chloride is less soluble than barium chloride; the solubility of the nitrates in water is approximately the same.

The salts of radium are the source of a spontaneous and continuous release of heat.

Pure radium chloride is paramagnetic. Its specific magnetization coefficient K (ratio of the magnetic moment of the unit of mass to the intensity of the field) was measured by MM. P. Curie and C. Chéneveau using a device established by these two physicists[10]. This coefficient was measured by comparison with that of water and corrected for the action of air magnetism. So we found

$$K = 1.05 \times 10^{-6}. \tag{1}$$

Pure barium chloride is diamagnetic, its specific magnetization coefficient is

$$K = -0.4 \times 10^{-6}. \tag{2}$$

Moreover, according to the previous results, it is found that a barium chloride containing radium, with approximately 17 per cent of radium chloride, is diamagnetic and has a specific coefficient[11]

$$K = 0.2 \times 10^{-6}. \tag{3}$$

Fractionation of Ordinary Barium Chloride.

We have sought to ascertain whether commercial barium chloride contains small quantities of radium chloride, which cannot be detected by our measuring device. To do this we fractionated a

---

[10] *Société de Physique*, 3 April 1903.

[11] In 1899, M. St. Meyer announced that the barium carbonate containing radium was paramagnetic (*Wied. Ann.*, Vol. LXVIII). However, M. Meyer had operated with a product with very little content of radium, and probably containing only $\frac{1}{1000}$ radium salts. This product should have been diamagnetic. It is likely that this body contained a small iron impurity.

great quantity of commercial barium chloride, hoping to concentrate by this process the trace of radium chloride if such were present.

Fifty kilos, of commercial barium chloride were dissolved in water; the solution was precipitated by hydrochloric acid free from sulphuric acid, which yielded 20 kilos, of the precipitated chloride. This was dissolved in water and partially precipitated by hydrochloric acid, which gave 8.5 kilos, of precipitated chloride. This chloride was fractionated by the method used for the barium chloride containing radium; and at the end of the process, 10 grms. of chloride were obtained, corresponding to the least soluble part. This chloride did not show any radioactivity in our measuring device; it therefore did not contain radium; this substance is, therefore, absent from the ores of barium.

Chapter III. Radiation of the New Radioactive Substances.

Methods of Investigation of the Radiation.

To study the radiation emitted by radioactive substances, we can use any of the properties of this radiation. We can therefore use either the action of the rays on photographic plates, or their property of ionising air and making it conductive, or their ability to cause the fluorescence of certain substances. Henceforth, in speaking of these different methods of working, I shall use, to abbreviate, the expressions radiographic method, electrical method, fluoroscopic method.

The first two were used from the beginning for the study of uranium rays; the fluoroscopic method can only be applied to new, highly radioactive substances, since weakly radioactive bodies such as uranium and thorium produce no appreciable fluorescence. The electrical method is the only one that produces precise intensity measures; the other two are specially apt to give qualitative results from this point of view, and can only provide rough intensity measurements. The results obtained with the three methods just considered are not strictly comparable with one another and may not be comparable at all. The sensitive plate, the gas which is ionised, the fluorescent screen, are all receivers, which absorb the energy of the radiation, and transform it into another form of energy: chemical energy, ionic energy, or luminous energy. Each receiver absorbs a fraction of the radiation, which essentially depends on its nature. Later on, we shall see that the radiation is complex, that the portions of the radiation absorbed by the different receivers may differ from each other both quantitatively and qualitatively. Finally, it is neither obvious, nor even probable, that the energy absorbed is entirely transformed by the receiver into the form that we wish to observe; part of this energy may be transformed into heat, into the emission of secondary radiations which, according to the case, may or may not assist in the production of the observed phenomenon, into chemical action differ from that which one observes, &c., and here again, the effective action of the receiver, with reference to the end we have in view, depends essentially upon the nature of this receiver.

Let us compare two radioactive samples, one of which contains radium and the other polonium, and which are equally active in the condenser of the apparatus in Fig. 1. If each is covered with a thin sheet of aluminium, the second will appear considerably less active than the first, and it will be the same if they are placed under the same fluorescent screen, when the latter is thick enough, or when it is placed at a certain distance from the two radioactive substances.

Energy of Radiation.

Whatever research method is used, we always find that the radiation energy of the new radioactive substances is considerably greater than that of uranium and thorium. Thus, at short distance, a photographic plate is impressed, so to speak, instantly, whereas an exposure of twenty-four hours is necessary when operating with uranium and thorium. A fluorescent screen is brightly lit in contact with the new radioactive substances, whilst no trace of luminosity is seen with uranium and thorium. Finally, the ionising action on air is also considerably more intense, in the ratio of approximately 10^6. However, it is no longer possible to estimate the *total intensity of the radiation*, as in the case of uranium, by the electrical method described at the beginning (Fig. 1). Indeed, with uranium, for example, the radiation is almost completely absorbed by the layer of air which separates the plates, and the limiting current is reached for a voltage of 100 volts. But this is no longer the case for highly radioactive substances. Part of the radium radiation consists of very penetrating rays, which pass through the condenser and the metallic plates, and are in no way utilised to ionise the air between the plates. In addition, the limiting current cannot always be obtained for the voltages available; for example, with very active polonium the current remains proportional to the tension between 100 and 500 volts. Therefore the experimental conditions which give a simple interpretation are not realised, and, consequently, the numbers obtained cannot be considered as giving the measurement of the total radiation; from this point of view, they are only a rough approximation.

Complex Nature of the Radiation.

The work of various physicists (MM. Becquerel, Meyer and von Schweidler, Giesel, Villard, Rutherford, M. and Mdme. Curie) have shown that the radiation of radioactive substances is very complex. It is advisable to distinguish three kinds of rays, which I shall designate, according to the notation adopted by Mr. Rutherford, by the letters α, β, γ.

I. The α-rays are very slightly penetrating, which seem to constitute the largest part of the radiation. These rays are characterised by the laws according to which they are absorbed by matter. The magnetic field acts very weakly upon them, and they were formerly considered as insensitive to the action of this field. However, in an intense magnetic field, the α-rays are slightly deflected; the deflection occurs in the same way as in the case of cathode rays, but the direction of the deflection is reversed; it is the same as for the canal rays of the Crookes tubes.

II. The β-rays are less absorbable as a whole than the preceding ones. They are deflected by a magnetic field in the same way and direction as the cathode rays.

III. The γ-rays are penetrating rays, unaffected by the magnetic field, and comparable to Röntgen rays.

The rays of the same group can have a penetrating power which varies within very wide limits, as has been proven for β-rays.

Consider the following imaginary experiment: — Some radium, R, is placed at the bottom of a small deep cavity, dug in a block of lead, P (Fig. 4).

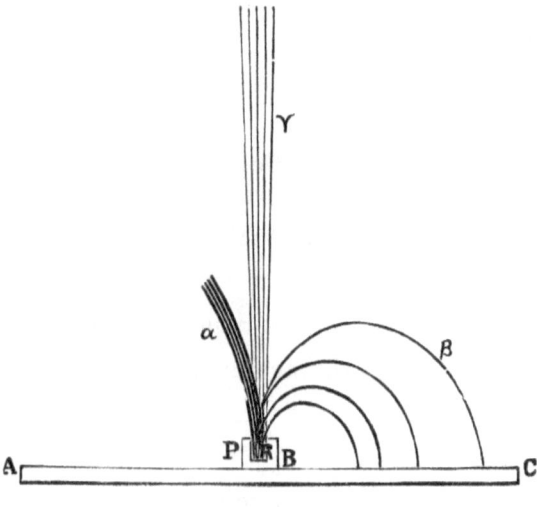

Figure 4.

A beam of rectilinear rays, slightly expanded, escapes from the receptacle. Suppose that, in the region surrounding the receptacle, a uniform, very strong magnetic field is established normal to the plane of the figure and directed towards the back of this plane.

The three groups of rays, α, β, γ, will now be separated. Then weak γ-rays continue their straight path without a trace of deviation. The β-rays are deflected like cathode rays, and describe circular trajectories in the plane of the figure whose radius varies within wide limits. If the receptacle is placed on a photographic plate, A C, the portion, B C, of the plate which receives the β-rays is impressed. Finally, the α-rays form a very intense beam which is slightly deflected, and which is absorbed fairly quickly by the air. These rays describe, in the plane of the figure, a trajectory whose radius of curvature is very large, the direction of the deflection being the opposite of that which takes place for the β-rays.

If we cover the receptacle with a thin sheet of aluminium (0.1 m.m. thick), the α-rays are suppressed almost entirely, the β-rays are much less, and the γ-rays do not appear to be absorbed significantly.

The experiment which I have just described was not carried out in this form, and one will see in the continuation which are the experiments which show the action of the magnetic field on the various groups of rays.

Action of the Magnetic Field.

We have seen that the rays emitted by radioactive substances have a large number of properties common to cathode rays and to Röntgen rays. Both cathode rays and Röntgen rays ionise the air, act on photographic plates, cause fluorescence, do not experience regular reflection. But the cathode rays differ from Röntgen rays in that they are deflected from their rectilinear path by the action of the magnetic field, and in that they carry negative charges of electricity.

The fact that the magnetic field acts on the rays emitted by radioactive substances was discovered almost simultaneously by MM. Giesel, Meyer and von Schweidler, and Becquerel[1]. These physicists have recognized that the rays of radioactive substances are deflected by the magnetic field in the same way and direction as cathode rays; their observations were related to the β-rays.

M. Curie has shown that the radiation of radium comprises two very distinct groups of rays, one of which is easily deflected by the magnetic field (β-rays), whilst the other remains insensitive to the action of this field[2] (α and γ-rays).

M. Becquerel did not observe any cathode ray emission from the polonium samples prepared by us. On the contrary, it was on a polonium sample, prepared by himself, that M. Giesel first observed the effect of the magnetic field. None of the polonium samples prepared by us have ever produced a cathode ray like emission.

The polonium of M. Giesel only gives rise to these rays when recently prepared, and it is likely that the emission is due to the phenomenon of induced radioactivity discussed below.

The following are experiments which prove that part of the radiation of radium, and one portion only, consists of easily deflected rays (β-rays). These experiments were done according to the electrical method[3].

The radioactive body A (Fig. 5) sends radiation in the direction AD between the plates P and P'. The plate P is maintained at the potential of 500 volts, plate P' is connected to an electrometer and

---

[1]Giesel, *Wied. Ann.*, 2 November 1899. — Meyer and von Schweidler, *Acad. Anzeiger Wien*, 3 and 9 November 1899. — Becquerel, *Comptes rendus*, 11 December 1899.

[2]P. Curie, *Comptes rendus*, 8 January 1900.

[3]P. Curie, *Comptes rendus*, 8 January 1900.

to a quartz electric piezometer. We measure the intensity of the current passing through the air under the influence of radiation. The magnetic field can be established at will perpendicular to the plane of the figure over the whole region *EEEE*. If the rays are deflected, even slightly, they no longer pass between the plates, and the current is suppressed. The region through which the rays pass is surrounded by the lead masses B, B', B", and by the armatures of the electro-magnet; when the rays are deflected, they are absorbed by the masses of lead B and B'.

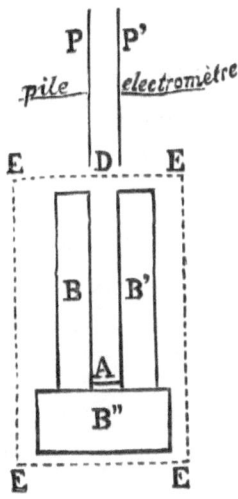

Figure 5.

The results obtained depend essentially on the distance, AD, of the radiating substance, A, from the condenser at D. If the distance AD is large enough (greater than 7 c.m.), the largest part of radium rays (90 to 100 per cent) arriving at the condenser are deflected and suppressed for a field of 2500 units. These are the β-rays. If the distance AD is less than 65 m.m., a smaller part of the rays are deflected by the action of the field; this part is also completely deflected by a field of 2500 units, and the proportion of rays suppressed is not increased when the filed is increased from 2500 to 7000 units.

The proportion of the rays not suppressed by the field increases with decreasing distance, AD, between the radiating body and the condenser. For short distances, the rays which can be easily deflected only constitute a very small fraction of the total radiation. The penetrating rays are therefore, for the most part, deviating rays of the cathode type (β-rays).

With the experimental device which has just been described, the action of the magnetic field on the α-rays could hardly be observed for the fields employed. The very important radiation, apparently not deviable, observed at a short distance from the radiating source, consisted of α-rays; the undeflected radiation observed at long distance consisted of γ-rays.

When the beam is sieved through an absorbing plate (aluminium or black paper), the rays which pass are almost all deflected by the field, so that using the screen and the magnetic field, produces an almost complete suppression of the radiation in the condenser, the remainder being due to the γ-rays, the proportion of which is small. The α-rays are absorbed by the screen.

An aluminium plate of 1/100 m.m. thickness is enough to remove almost all the rays not readily deflected when the substance is far enough from the condenser; for smaller distances (34 m.m. and 51 m.m.) two aluminium sheets are necessary to achieve the same result.

Similar measurements were made on four substances containing radium (chlorides or carbonates) of very different activity; the results obtained were very similar.

It can be noted that, for all the samples, the penetrating rays deflected by the magnet (β-rays) are only a small part of the total radiation; they only play a small part in measurements where the whole radiation is used to produce the conductivity of the air.

The radiation emitted by polonium may be studied by the electrical method. When we vary the distance, *AD*, of the polonium from the condenser, we first observe no current as long as the distance is large enough; when we bring the polonium closer, we observe that, for a certain distance which was 4cm for the sample studied, the radiation is very suddenly felt with a fairly great intensity; the current then increases regularly if we continue to bring the polonium closer, but the magnetic field does not produce an appreciable effect under these conditions. It seems that the radiation of polonium is delimited in space and hardly passes in the air beyond a kind of sheath surrounding the substance to a thickness of a few centimeters.

There are important general caveats about the significance of the experiments I have just described. When I indicate the proportion of rays deflected by the magnet, it refers only to radiation that can activate a current in the capacitor. In employing the fluorescent action of the Becquerel rays, the proportion would probably be different, an intensity measurement generally having meaning only for the measurement method used.

The rays of polonium are α-rays. In the experiments I have just described, no effect of the magnetic field was observed on these rays, but the experimental set-up was such that a small deflection would pass unnoticed.

Experiments by the radiographic method have confirmed the preceding results. By using radium as a radiant source, and receiving the impression on a plate parallel to the primitive beam and normal to the field, we obtain the very clear trace of two beams separated by the action of the field, one deflected, the another not deflected. The β-rays constitute the deflected beam; the rays α, being slightly deflected, substantially merge with the non-deflected beam of the γ-rays.

Deflected β-Rays.

The experiments of M. Giesel and MM. Meyer and von Schweidler showed that the radiation of the radioactive bodies is, at least partly, deflected by a magnetic field, and that the deflection resembles that of the cathode rays. M. Becquerel studied the action of the field on the rays by the radiographic method[4]. The experimental device used was that of Fig. 4. The radium was placed in the lead container, P, and this receptacle was placed on the sensitive face of a photographic plate, AC, wrapped in black paper. The whole was placed between the poles of an electro-magnet, the magnetic field being normal to the plane of the figure.

If the field is directed towards the back of this plane, the part BC of the plate is impressed by rays which, having described circular trajectories, return back to the plate and strike it at a right angle. These rays are β-rays.

M. Becquerel has shown that the impression consists of a broad diffused band, a true continuous spectrum, showing that the beam of deviable rays emitted by the source consists of an infinity of unevenly deflectable radiation. If we cover the gelatin of the plate with various absorbent screens (paper, glass, metals), one portion of the spectrum is suppressed, and we see that the rays most deflected by the magnetic field — in other words those which have the smallest radius of curvature — are the most strongly absorbed. For each screen, the impression on the plate begins at a certain distance from the radiant source, this distance being proportional to the absorptive power of the screen.

---

[4]Becquerel, *Comptes rendus*, Vol. CXXX, p. 206, 372, 810.

Charge of the Deflected Rays.

The cathode rays are, as shown by M. Perrin, charged with negative electricity[5]. Further, they can, according to the experiments of M. Perrin and M. Lenard[6], transport their charge through metallic envelopes connected to the ground and through isolating screens. At every point where the cathode rays are absorbed, there is a continuous release of negative electricity. We have have found that the same is true for the deflected β-rays of radium. *The deviable β-rays of radium are charged with negative electricity[7].*

(Note. — Let the radioactive substance be placed on one of the plates of a condenser, this plate being connected to earth; the second plate is connected to an electrometer, it receives and absorbs the rays emitted by the substance. If the rays are charged, there should be a continuous flow of electricity into the electrometer. This experiment, carried out in air, did not allow us to detect a charge accompanying the rays, but such an experiment is not sensitive. The air between the plates being made conductive by the rays, the electrometer is no longer isolated, and can only respond to sufficiently strong charges. In order that the α-rays may not interfere with the experiment, they can be suppressed by covering the source of radiation with a thin metallic screen; the results of the experiment is not modified.[8] We repeated this experiment, without more success, by letting the rays pass through the interior

---

[5] *Comptes rendus*, Vol. CXXI, p. 1130. *Annales de Chimie et de Physique*, Vol. II, 1897.

[6] Lenard, *Wied. Ann.*, Vol. LXIV, p. 279.

[7] M. and Mme Curie, *Comptes rendus*, 5 March 1900.

[8] In fact, in these experiments, there is always a deviation in the electrometer, but it's easy to see that the displacement is an effect of the electromotive contact force which exists between the plate connected to the electrometer and the neighboring conductors; this electromotive force deflects the electrometer, thanks to the conductivity of air exposed to radium radiation.

of a Faraday cylinder in connection with the electrometer[9].

One can already see from the above experiments that the charge of the rays of the radiating body employed was a weak one.

To note a weak release of electricity on the conductor which absorbs the rays, it is necessary that this conductor is well insulated; to obtain this result, it is necessary to screen it from air, either by placing it in a tube with a very perfect vacuum, or by surrounding it with a good solid dielectric. We employed the latter arrangement.

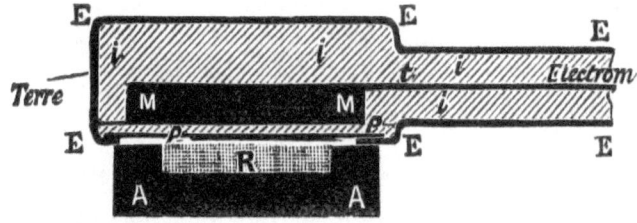

Figure 6.

A conducting disc, MM (Fig. 6), is connected by the wire, t, to the electrometer; the disc and wire are completely surrounded by the insulating material $iiii$; the whole is again covered with the metallic envelope, $EEEE$, which is in electric connection with the earth. On one side of the disc, the insulator, pp, and the metallic envelope are very thin. It is this face that is exposed to the radiation of the barium and radium salt, R, placed outside in a lead receptacle[10]. The rays emitted by the radium penetrate the

---

[9]The Faraday cylinder is not necessary, but it could have some advantages in the event that a strong diffusion of rays occurs when they struck the walls. We could thus hope to collect and use these scattered rays, if there are any.

[10]The insulating envelope must be perfectly smooth. Any air-filled cracks from the inner conductor to the metallic enclosure can cause a current due to the electromotive forces of contact that develop as a consequence of the conductivity of air under the action of radium.

metallic envelope and the insulating strip, *pp*, and are absorbed by the metallic disc, *MM*. This is then the source of a continuous and constant release of negative electricity, which can be seen on the electrometer, and which is measured using piezoelectric quartz.

The current thus created is very weak. With very active barium-radium chloride, forming a layer of 2.5 sq. c.m. of surface, and of 0.2 c.m. in thickness, a current of the order of magnitude 10^{-11} amperes is obtained, the rays used having crossed, before being absorbed by the disc M M, a thickness of aluminium of 0.01 m.m., and a thickness of ebonite of 0.3 m.m.

We have successively used lead, copper, and zinc for the disc *MM*, ebonite and paraffin for the insulation; the results obtained were the same.

The current decreases when the radiant source *R* is moved away, or when a less active product is used.

We obtained the same results again by replacing the disc *MM* with a Faraday cylinder filled with air, but wrapped externally with insulating material. The opening of the cylinder, closed by the thin insulating plate, *pp*, was then opposite the radiating source.

Finally, we did the opposite experiment, which consists in placing the lead receptacle with the radium in the centre of the insulating material and in connection with the electrometer (Fig. 7), the whole being enveloped by the metallic enclosure connected to earth.

Under these conditions, it is observed on the electrometer that the radium has a positive charge equal in magnitude to the negative charge of the former experiment. The radium rays penetrate the thin dielectric plate, *pp*, and leave the conductor inside carrying with them negative electricity.

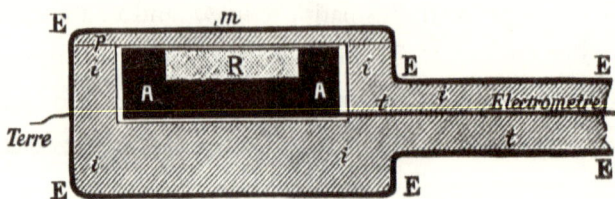

Figure 7.

The α-rays of radium do not interfere in these experiments, being absorbed almost completely by a very thin layer of material. The method which has just been described is also not suitable for the study of the charge of the polonium rays, these rays being also very little penetrating. We observed no indication of any charge in the case of polonium, which emits only α-rays; but, for the above reason, no conclusion can be drawn from this.

Thus, in the case of the deflected β-rays of radium, as in the case of cathode rays, the rays transport electricity. However, so far we have never recognized the existence of electric charges unrelated to matter. In the study of the emission of the β-rays of radium, we are therefore led to make use of the same theory as that currently in use for the study of cathode rays. In this ballistic theory, formulated by Sir William Crookes, then developed and completed by Prof. J. J. Thomson, the cathode rays consist of extremely minute particles, which are launched from the cathode with a very high speed, and which are charged with negative electricity. We can likewise imagine that radium sends into space negatively charged particles.

A sample of radium, enclosed in a solid, thin, perfectly insulated envelope, must spontaneously charge to a very high potential. By the ballistic hypothesis the potential would increase until the potential difference with the surrounding conductors became sufficient to hinder the ejection of the electrified particles and to cause their return to the source of radiation.

We have accidentally performed the experiment in question here. A sample of very active radium had long been enclosed in a glass vessel. In order to open the vessel, we made a line on the glass with a glass knife. At that moment, we clearly heard the noise of a spark, and then examining the vessel with a magnifying glass, we noticed that the glass had been perforated by a spark at the spot where it had been weakened by the scratch. The phenomenon produced is exactly comparable to the rupture of the glass of an overcharged Leyden jar.

The same phenomenon occurred with another glass. In addition, the moment the spark exploded, M. Curie, who was holding the glass, felt the electric shock from the discharge in his fingers.

Some glasses have good insulating properties. If you put the radium in a sealed, well-insulated glass vessel, you can expect that, at a given moment, the vessel will be spontaneously pierced.

Radium is the first example of a body that spontaneously charges with electricity.

Action of the Electric Field on the Deflected β-Rays of Radium.

The β-rays of radium, being analogous to the cathode rays, must be deflected by an electric field in the same way as the latter; i.e., as would a particle of matter negatively charged and launched into space with high speed. The existence of this deflection has been demonstrated both by M. Dorn[11] and M. Becquerel[12].

Consider a ray that crosses the space between the two plates

---

[11]Dorn, *Abh. Halle*, March 1900.
[12]Becquerel, *Comptes rendus*, Vol. CXXX, p. 819.

of a condenser. Suppose the direction of the ray parallel to the plates: when an electric field is established between them, the ray is subjected to the action of this uniform field over the entire length of the path in the condenser *l*. By virtue of this action the ray is deflected towards the positive plate and describes a parabolic arc; on leaving the field, it continues its path in a straight line, following the tangent to the arc of the parabola at the exit point. The ray can be received on a photographic plate perpendicular to its original direction. We observe the impression produced on the plate when the field is zero, and when it has a known value, and we deduce from there the value of the deflection, δ, which is the distance of the points in which the new direction of the ray and its original direction meet a common plane perpendicular to the original direction. If *h* is the distance of this plane from the condenser, i.e., at the edge of the field, we have, by a simple calculation, —

$$\delta = \frac{eFl\left(\frac{l}{2} + h\right)}{mv^2};$$

m being the mass of the moving particles, *e* its charge, *v* its velocity, and *F* the strength of the field.

The experiments of M. Becquerel allowed him to give an approximate value to δ.

Ratio of Charge to Mass for a Particle Negatively Charged Emitted by Radium.

When a material particle having a mass *m* and a negative charge *e*, is launched with a velocity *v* into a uniform magnetic field normal to its initial velocity, this particle describes, in a plane normal to the field and passing through its initial velocity, an arc of a circle of radius ρ, such that — *H* being the strength of the field — we have the relation —

$$H\rho = \frac{m}{e}v.$$

If one measured for the same ray the deflection, δ, and the radius of curvature, ρ, in a magnetic field, one will be able, from these two experiments to extract the values of the ratio $\frac{e}{m}$ and velocity, v.

The experiments of M. Becquerel provided a first indication on this subject. They gave for the ratio $\frac{e}{m}$ a value approximately equal to 10^7 absolute electro-magnetic units, and for v a magnitude of 1.6×10^{10}. These values are of the same order of magnitude as for cathode rays.

Precise experiments have been made on the same subject by M. Kaufmann[13]. This physicist subjected a very narrow beam of to the simultaneous action of an electric field and a magnetic field, the two fields being uniform and having the same direction, normal to the original direction of the beam. The impression produced on a plate normal to the original beam and placed above the limits of the field relative to the source, takes the form of a curve, each point of which corresponds to one of the rays of the original heterogeneous beam. The most penetrating and least deflected rays are those with the greatest velocity.

It follows from the experiments of M. Kaufmann, that for radium rays, whose speed is notably higher than that of cathode rays, the ratio $\frac{e}{m}$ decreases as the velocity increases.

According to the work of J. J. Thomson and Townsend[14], we may assume that the moving particle, which constitutes the ray, has a charge, e equal to that carried by an atom of hydrogen in electrolysis, this charge being the same for all the rays. We

---

[13]Kaufmann, *Nachrichten d. k. Gesell. d. Wiss. zu Goettingen*, 1901, Heft 2.
[14]Thomson, *Phil. Mag.*, Vol. XLVI, 1898. — Townsend, *Phil. Trans.* Vol. CXCV, 1901.

are therefore led to conclude that the mass of the particle, m, increases with increasing velocity.

However, theoretical considerations lead us to conceive that the inertia of the particle is precisely due to its state of charge in motion, the velocity of an electric charge in motion being incapable of modification without energy expenditure. In other words, the inertia of the particle is of electro-magnetic origin, and the mass of the particle is— at least partly — a virtual mass or an electro-magnetic mass. M. Abraham[15] goes further, and assumes that the mass of the particle is entirely an electro-magnetic mass. If, according to this hypothesis, we calculate the value of this mass, m, for a known velocity, v, we find that m approaches infinity when v approaches the speed of light, and that m approaches a constant value when the velocity, v, is much lower than that of light. The experiments of M. Kaufmann are in agreement with the results of this theory, which is very important since it allows to foresee the possibility of establishing the bases of mechanics on the dynamics of small particles of matter charged in a state of motion[16].

These are the numbers obtained by M. Kaufmann for $\frac{e}{m}$ and v.

M. Kaufmann concludes, from the comparison of his experiments with theory, that the limiting value of the ratio $\frac{e}{m}$ for relatively low speed radium rays would be the same as the value $\frac{e}{m}$ for cathode rays.

The most complete experiments of M. Kaufmann were made with a tiny quantity of pure radium chloride, which we have

---

[15]Abraham, *Nachrichten d. k. Gesell. d. Wiss. zu Goettingen*, 1902, Heft 1.

[16]Some developments on this issue as well as a complete study of charged materials (electrons or corpuscles) and the references of the related works are in the Thesis of Mr. Langevin.

$\frac{e}{m}$		$v \frac{\text{c.m.}}{\text{sec.}}$		
Electro-magnetic units.				
1.865	$\times 10^7$	0.7	$\times 10^{10}$	For cathode rays (Simon).
1.31	$\times 10^7$	2.36	$\times 10^{10}$	For radium rays (Kaufmann).
1.17	$\times 10^7$	2.48	$\times 10^{10}$	
0.97	$\times 10^7$	2.59	$\times 10^{10}$	
0.77	$\times 10^7$	2.72	$\times 10^{10}$	
0.63	$\times 10^7$	2.83	$\times 10^{10}$	

placed at his disposal.

According to M. Kaufmann's experiments, some β-rays of radium have a speed very near to that of light. We understand that these so fast rays can possess a very penetrating power vis-à-vis matter.

Action of the Magnetic Field on α-Rays.

In a recent work, Mr. Rutherford announced[17] that, in a strong electric or magnetic field, the α-rays of radium are slightly deflected like positively electrified particles possessing great speed. Mr. Rutherford concludes from his experiments that the velocity of the α-rays is of the order of magnitude $2.5 \times 10^9 \frac{\text{c.m.}}{\text{sec.}}$ and that the ratio $\frac{e}{m}$ for these rays is of the order of magnitude 6×10^3, which is 10^4 times greater than for the deflected β-rays. We will see later that these conclusions of Mr. Rutherford are in agreement with the previously known properties of the α-radiation, and that they account, at least in part, for the law of absorption of this radiation.

The experiments of Mr. Rutherford have been confirmed by M. Bacquerel. M. Becquerel has shown, moreover, that polonium

---

[17]Rutherford, *Physik. Zeitschrift*, 15 January 1908.

rays behave in a magnetic field like the α-rays of radium, and that, for the same field, they seem to have the same curvature as the latter.

It also follows from M. Becquerel's experiments that the α-rays do not form a magnetic spectrum, but rather behave like a homogeneous radiation, all the rays being equally deflected[18].

M. Des Coudres made a measurement of the electrical deflection and the magnetic deflection of the α-rays of radium in a vacuum. He found for the speed of these rays $v = 1.65 \times 10^9 \frac{cm}{sec}$ and for the ratio of charge to mass $\frac{e}{m} = 6400$ electromagnetic units[19]. The speed of α-rays is therefore about 20 times lower than that of light. The ratio $\frac{e}{m}$ is of the same order of magnitude as that found for the hydrogen in electrolysis: $\frac{e}{m} = 9650$. If we assume that the charge of each projectile is the same as that of a hydrogen atom in electrolysis, we conclude that the mass of this projectile is of the same order of magnitude as that of an hydrogen atom.

Now we have just seen that, for the slowest cathode rays and for the β rays of radium, the ratio $\frac{e}{m}$ is 1.865×10^7; this ratio is therefore approximately 2000 times greater than that obtained in electrolysis. Since the charge of the negatively charged particle is assumed to be the same as that of an hydrogen atom, its limit mass at relatively low speeds would therefore be approximately 2000 times smaller than that of a hydrogen atom.

The projectiles which constitute the β-rays are therefore both much smaller and possess a greater speed than those which constitute the α rays. It is therefore easy to understand that the former have a much greater penetrating power than the latter.

---

[18]Becquerel, *Comptes rendus*, of the 26 January and 16 February 1900.
[19]Des Coudres, *Physik. Zeitschrift.*, 1st June 1903.

Action of the Magnetic Field on the Rays of other Radioactive Substances.

We have just seen that radium gives off α-rays comparable to the tube rays, β-rays comparable to cathode rays, and γ-rays which are penetrating and not deflected. Polonium emits only α-rays. Among other radioactive substances, actinium appears to behave like radium, but the study of its radiation is not yet as advanced as that of the radium radiation. As for weakly radioactive substances, we now know that uranium and thorium emit both α-rays and deflectable β-rays (Becquerel, Rutherford).

Proportion of Deflectable β-Rays in the Radiation of Radium.

As I have already mentioned, the proportion of β-rays increases as we move away from the source of radiation. However, these rays never occur alone, and for long distances we always observe the presence of γ-rays. The presence of highly penetrating, un-deflected rays in the radiation of radium was, for the first time, observed by M. Villard[20]. These rays constitute only a small part of the radiation measured by the electrical method, and their presence had escaped us in our first experiments, so that we then wrongly believed that the radiation at great distances contained only deflectable rays.

The following are the numerical results obtained with experiments made by the electrical method with an apparatus similar to that of Fig. 5. The radium was only separated from the condenser by ambient air. I denote by the letter d the distance from the source of radiation to the condenser. Assuming equal to 100 the

---

[20]Villard, *Comptes rendus*, Vol. CXXX, p. 1010.

current obtained without magnetic field for each distance, the numbers of the second line indicate the current subsisting when the magnetic field is acting. These numbers can be considered as giving the percentage of all the α– and γ-rays, the deflection of the α-rays having been hardly observable with the device used.

At great distances there are no α-rays, and the undeflected radiation is therefore of the γ type only.

Experiments made at short distances: —

d, in centimeters	3.4	5.1	6.0	6.5
Percentage of undeflected rays	74	56	33	11

Experiments made at long distances, with a product considerably more active than that which had been used for the previous series: —

d, in centimeters	14	30	53	80	98
Percentage of undeflected rays	12	14	17	14	16
d, in centimeters	124	157			
Percentage of undeflected rays	14	11			

We see that, from a certain distance, the proportion of undeflected rays in the radiation is approximately constant. These rays probably all belong to the γ species. There is no need to take too much trouble to account for the irregularities in the numbers of the second line, if we consider that the total intensity of the current in the two extreme experiments was in the ratio of 660 to 10. The measurements could have continued to a distance of 1,57 m. from the radiant source, and we could go even further now.

The following is another series of experiments in which the radium was enclosed in a very narrow glass tube, placed below the condenser and parallel to the plates. The rays emitted passed through a certain thickness of glass and air before entering the condenser: —

d, in centimeters	2.5	3.3	4.1	5.9	7.5	9.6
Percentage of undeflected rays	33	33	21	16	14	10

d, in centimeters	11.3	13.9	17.2
Percentage of undeflected rays	9	9	10

As in the preceding experiments, the number of the second line tend towards a constant value, when the distance d increases, but the limit is reached for smaller distances than in the previous series, because the α-rays have been more strongly absorbed by the glass than the β- and γ-rays.

The following experiment shows that a thin sheet of aluminium (0.01 m.m. thick) absorbs mainly α-rays. The product being placed 5 c.m. from the condenser, the proportion of rays other than β, when the magnetic field is acting, is about 71 per cent. When the same substance is covered with the sheet of aluminium, and the distance remaining the same, we find that the transmitted radiation is almost totally deflected by the magnetic field, the α-rays having been absorbed by the sheet. The same result is obtained when paper is used as an absorbent screen.

Most of the radiation of radium consists of α-rays, which are probably emitted mainly by the surface layer of the radiating material. When the thickness of the layer of radiating matter is varied, the intensity of the current increases with this thickness; the increase is not proportional to the thickness for all of the radiation; it is, moreover, more noticeable for the β-rays than for the α-rays, so that the proportion of the β-rays increases with the thickness of the active layer. The radiant source being placed at a distance of 5 c.m. from the condenser, we find that, for a thickness equal to 0.4 m.m. of the active layer, the total radiation is given by the number 28, and the proportion of the β-rays is 29 per cent. By making the layer 2 m.m. thick, i.e., five times greater, we obtain a total radiation equal to 102, and a proportion of deflecting β-rays equal to 45 per cent. The total radiation which remains at this distance has therefore been increased in the ratio of 3.6, and the deflecting β-radiation has become approximately five times stronger.

The previous experiments were made by the electrical method. When the radiographic method is used, some results seem to be in contradiction with the above. In the experiments of M. Villard, a beam of radium rays, subjected to the action of the magnetic field, was received on a stack of photographic plates. The undeflected and penetrating beam γ passed through all the plates, and left its mark on each of them. The deflected beam β produced an impression on the first plate only. This beam appeared therefore to contain no rays of great penetration.

On the contrary, in our experiments a beam which propagates in the air contains at the greatest distances accessible to the observation about 9/10 of γ-rays, and the same is the case when the source of radiation is enclosed in a small sealed glass vessel. In M. Villard's experiments, these deflected and penetrating β-rays did not impress the photographic plates placed beyond the first, because they are largely diffused in all directions by the first solid obstacle encountered and cease to form a beam. In our experiments, the rays emitted by radium and transmitted through the glass of the vessel were also probably scattered by the glass, but the vessel being very small then functioned itself as a source of deflecting β-rays at its surface, and we were able to observe these up to a great distance from the vessel.

The cathode rays of Crookes tubes can only pass through very thin screens (aluminium screens up to 0.01 m.m. thick). A beam of rays striking the screen normally is scattered in all directions; but the diffusion becomes less important as the screen is thinner, and for very thin screens the emerging beam is practically the extension of the incident beam[21].

The deflected β-rays of radium behave in a similar manner, but the transmitted beam experiences, for an equal screen thickness, a much less profound modification. According to the experiments of M. Becquerel, the very highly deflectable β-rays of radium (those whose speed is relatively low) are strongly scattered by an aluminium screen of thickness 0.1 m.m.; but the penetrating and less deflected rays (rays of the cathode kind of great velocity) cross this same screen without any significant diffusion, regardless of the inclination of the screen with respect to the direction of the beam. High speed β-rays penetrate through a much greater thickness of paraffin (several centimetres) without diffusion, and in this the curvature of the beam produced by the magnetic field can be traced. The thicker the screen, and the more absorbent its material, the more the original deflectable beam is altered, because, as the thickness of the screen increases, the diffusion occurs progressively among new groups of rays more and more penetrating.

The β-rays of radium experience a diffusion in passing through the air, which is very marked for readily deflected rays, but which is much slighter than that produced by equal thicknesses of solid substances. This is why the deflectable β-rays propagate in the air at great distances.

---

[21]Des Coudres, *Physik. Zeitschrift*, November 1902.

Penetrating Power of the Radiation of Radioactive Bodies.

Since the beginning of the researches on radioactive bodies, we were interested in the absorption produced by different screens upon the rays emitted by these bodies. In a first paper on this subject[22] I gave numbers (quoted at the beginning of this work) indicating the penetrating power of uranium and thorium rays. Mr. Rutherford has studied uranium radiation[23], and proved that it is heterogeneous. Mr. Owens concluded the same for thorium rays[24]. When the discovery of strongly radioactive bodies immediately followed upon this, the penetrating power of their rays was also studied by various physicists (Becquerel, Meyer and von Schweidler, Curie, Rutherford). The first observations highlighted the complexity of the radiation, which seems to be a general phenomenon common to radioactive substances[25]. We are here in the presence of sources, which emit a variety of radiations, each of which has its own penetration power.

The question is further complicated by this fact, that it is necessary to investigate to what extent the nature of the radiation can be modified by the passage through material substances and that, consequently, each set of measures has precise meaning only for the experimental device used. Having outlined these caveats, we can try to coordinate the various experiments and to present all of the results obtained.

Radioactive bodies emit rays which propagate both in the air and *in vacuo*. The propagation is rectilinear; this fact is proved

---

[22]Mme Curie, *Comptes rendus*, April 1898.

[23]Rutherford, *Phil. Mag.*, January 1899.

[24]Owens, *Phil. Mag.*, October 1899.

[25]Becquerel, *Rapports au Congrés de Physique*, 1900. — Meyer and von Schweidler, *Comptes rendus de l'Acad. de Vienne*, March 1900 (*Physik. Zeitschrift*, Vol. I, p. 209).

by the sharpness and shape of the shadows formed by interposing bodies opaque to the radiation between the source and the sensitive plate or fluorescent screen which serves as receiver, the source having small dimensions compared to its distance from the receiver.

Various experiments which prove the rectilinear propagation of uranium, radium, and polonium rays were made by M. Becquerel[26].

It is interesting to know the distance at which the rays can propagate in air. We have found that radium emits rays that can be detected in air from several metres away. In some of our electrical measurements, the action of the source on the air of the condenser was exerted at a distance between 2 and 3 metres. We have also obtained fluorescent effects and radiographic impressions at similar distances. The experiments can only be done easily with very intense radioactive sources, because, independently of the absorption exerted by the air, the action on a given receiver varies inversely as the square of the distance from a source of small dimensions. This radiation, which propagates at a great distance in the case of radium, includes both cathode rays and rays which are undeflected; however, the deflected rays predominate, from the results of the experiments mentioned above. The greater part of the radiation (α-rays) is, on the contrary, limited in air to a distance of about 7 c.m. from the source.

I made some experiments with radium enclosed in a small glass vessel. The rays emerging from the vessel crossed a certain space of air and were received in a condenser, which was used to measure their ionising capacity by the ordinary electrical method. We varied the distance, d, from the source to the condenser, and

---

[26]Becquerel, *Comptes rendus*, Vol. CXXX, p. 979 and 1154.

measured the current of saturation, i, in the condenser. The following are the results of one of the series of determinations: —

d, c.m.	i.	$(i \times d^2) \times 10^{-3}$
10	127	13
20	38	15
30	17.4	16
40	10.5	17
50	6.9	17
60	47	17
70	3.8	19
100	1.65	17

After a certain distance, the intensity of radiation varies inversely as the square of the distance from the condenser.

The radiation of polonium only propagates in the air to a distance of a few centimetres (4 to 6 c.m.) from the source of radiation.

If we consider the absorption of radiation by solid screens, there is again a fundamental difference between radium and polonium. Radium emits rays capable of penetrating through a great thicknesses of solid matter, e.g., several centimetres of lead or of glass[27]. The rays which have passed through a great thickness of a solid body are extremely penetrating, and it is practically impossible to absorb them entirely by any material whatsoever. But these rays constitute only a small fraction of the total radiation, whose large mass is, on the contrary, absorbed by a small thickness of solid matter.

Polonium emits rays which are readily absorbed, and which can only pass through very thin solid screens.

---

[27]M. and Mme Curie, *Rapports au Congrés* 1900.

Here, by way of example, are some numbers showing the absorption produced by an aluminium lamina of thickness 0.01 m.m. This lamina was placed above and almost in contact with the substance. The direct radiation and that transmitted by the aluminium were measured by the electrical method (apparatus of Fig. 1); the saturation current was practically obtained in all cases. I have represented the activity of the radiating body by a, that of uranium being unity.

	a.	Fraction of radiation transmitted.
Chloride of barium and radium	57	0.32
Bromide of barium and radium	43	0.30
Chloride of barium and radium	1200	0.30
Sulphate of barium and radium	5000	0.29
Sulphate of barium and radium	10000	0.32
Metallic bismuth and polonium	—	0.22
Compounds of uranium	—	0.20
Compounds of thorium in a thin layer	—	0.38

We see that radium compounds of different nature and activity give very similar results, as I already indicated for the compounds of uranium and thorium at the beginning of this work. We see also that, taking into account the whole amount of radiation, and with a given absorbent screen, the various radioactive substances can be arranged in the following decreasing order of penetrating power: — Thorium, radium, polonium, uranium.

These results are similar to those published by Mr. Rutherford in a Memoir relating to this question[28].

Mr. Rutherford finds, moreover, that the order is the same when the absorbent substance is constituted by air. But it is likely that this order is not absolute, and would not be maintained inde-

---

[28]Rutherford, *Phil. Mag.*, July 1902.

pendently of the nature and thickness of the screen. Experience has shown, in fact, that the law of absorption is very different for polonium and radium, and that, for the latter, it is necessary to consider the absorption of the rays of each of the three groups separately.

Polonium is particularly well adapted to the study of α-rays, because the samples which we possess do not emit other kind of rays. I made a preliminary series of experiments with extremely active recently prepared specimens of polonium. I found the absorbability of the rays to increase with increase of thickness of the matter traversed[29]. This singular law of absorption is contrary to that known for other kinds of radiation.

For this study, I used our device for measuring electrical conductivity arranged in the following manner: —

The two plates of a condenser, PP and $P'P'$ (Fig 8), are horizontal and disposed in a metallic box, $BBBB$, connected to earth. The active body, A, located in a thick metallic box, $CC\,CC$, connected with the plate $P'P'$, acts on the air of the condenser across a metallic sheet, T; the rays which pass through the sheet are the only ones utilised for producing the current, the electric field being limited by the sheet. The distance, AT, of the active body from the sheet can be varied. The field between the plates is established by means of a battery; the current is measured using an electrometer and a piezoelectric quartz.

By placing various screens in A on the active body, and by adjusting the distance AT, we can measure the absorption of rays which travel long or short distances in the air.

The following are the results obtained with polonium: —

For a certain value of the distance AT (4 c.m. and more), no

---

[29]Mme Curie, *Comptes rendus*, 8 January 1900.

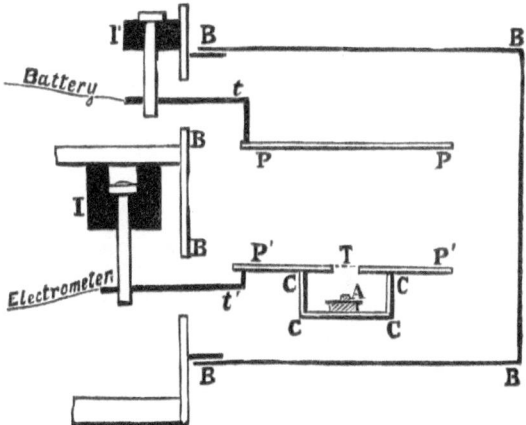

Figure 8.

current flows; the rays do not penetrate the condenser. When the distance AT is reduced, the appearance of the rays in the condenser is quite abrupt, so that, for a small decrease in the distance, we pass from a very weak current to one of considerable strength; the current then increases regularly as the active body continues to approach the sheet T.

When we cover the active body with a sheet of aluminium 1/100 m.m. thick, the absorption produced by the lamina becomes greater, the greater the distance AT.

If a second similar lamina of aluminium is placed on the first, each absorbs a fraction of the radiation it receives, and this fraction is greater for the second lamina than for the first, so that it is the second lamina that seems more absorbent.

In the following table I have represented in the first line the distances in centimetres between the polonium and the sheet T; in the second line the percentage of the rays transmitted by a sheet of aluminium; in the third line the percentage of the rays transmitted by two sheets of the same aluminium: —

Distance AT	3.5	2.5	1.9	1.45	0.5
Percentage of rays transmitted by one lamina	0	0	5	10	25
Percentage of rays transmitted by two laminae	0	0	0	0	0.7

In these experiments the distance of the plates, P and P', was 3 c.m. It can be seen that the interposition of the aluminium screen decreases the intensity of the radiation in a greater proportion at further distances than at nearer distances.

This effect is even more pronounced than the numbers above seem to indicate. For a distance of 0.5 c.m. 25 per cent represents the mean penetration for all the rays which pass beyond this distance. If, for example, only those rays between 0.5 c.m. and 1 c.m. be comprehended, the penetration would be greater. And if the plate P be placed at a distance of 0.5 c.m. from P' the fraction of the radiation transmitted by the aluminium lamina (for AT = 0.5 c.m.) is 47 per cent, and through two laminae it is 5 per cent of the original radiation.

I have recently performed a second series of experiments with these same polonium samples, the activity of which was considerably reduced, the time interval between the two series of experiments being three years.

In the past experiments, polonium was in the form of a nitrite; in recent ones, it was in the state of metallic grains, obtained by fusing the nitrite with potassium cyanide.

I found that the radiation of polonium had preserved its essential characteristics, and I discovered new results. Here, for different values of the distance AT, are the fractions of the radiation transmitted by a screen formed by four superimposed very thin leaves of beaten aluminium.

Distance AT	0	1.5	2.6
Percentage of rays transmitted by the screen	76	66	39

I also noticed that the fraction of radiation absorbed by a given screen increases with the thickness of material which has already been crossed by the radiation, but this only takes place from a certain value of the distance AT. When this distance is zero (the polonium being in contact with the sheet, outside or inside the condenser), we observe that, with several similar superimposed screens, each absorbs the same fraction of the radiation it receives; in other words, the intensity of the radiation decreases therefore according to an exponential law as a function of the thickness of the material traversed, as would occur for a homogeneous radiation transmitted by the lamina without changing its nature.

The following numerical results are given with reference to these experiments: —

For a distance AT equal to 1.5 c.m. a thin aluminium screen transmits the fraction 0.51 of the radiation it receives when acting alone, and the fraction 0.34 of the radiation it receives when it is preceded by another similar screen.

On the contrary, for a distance AT equal to zero, this same screen transmits in the two cases considered the same fraction of the radiation that it receives, and this fraction is equal to 0.71; it is therefore greater than in the previous case.

The following numbers indicate for a distance AT equal to 0 and for a succession of thin superposed screens, the ratio of the radiation transmitted to the radiation received for each screen: —

Series of nine very thin copper leaves.	Series of seven very thin aluminium leaves.
0.72	0.69
0.78	0.94
0.75	0.95
0.77	0.91
0.70	0.92
0.77	0.93
0.69	0.91
0.79	
0.68	

Given the difficulties of using very thin screens and the super-imposition of screens in contact, the numbers in each column can be considered constant; only the first number in the aluminium column indicates a higher absorption than that indicated by the following numbers.

The α-rays of radium behave like the rays of polonium. We can study these rays almost exclusively by deflecting to one side the β-rays with the magnetic field; the γ-rays seem, in fact, unimportant compared to α-rays. However, this can be done only at some distance from the source of radiation. The following are the results of an experiment of this kind. We measured the fraction of the radiation transmitted by a lamina of aluminium 0.01 m.m. thick; this screen was always placed in the same position, above and at a short distance from the radiant source. With the apparatus of Fig. 5, the current produced in the condenser for different values of the distance $A\,D$ is observed, both in the presence and absence of the screen: —

Distance AD	6.0	5.1	3.4
Percentage of rays transmitted by the aluminium	3	7	24

The rays which travel furthest in the air are those most absorbed by the aluminium. There is therefore a great analogy between the absorbable α-rays of radium and the rays of polonium.

The deflected β-rays and the undeflected penetrating γ-rays
are, on the contrary, of a different nature. The experiments of
various physicists, in particular MM. Meyer and von Schweidler[30],
clearly show that, if we consider all of the radiation of radium, the
penetrating power of this radiation increases with the thickness
of the material traversed, as occurs with Röntgen rays. In these
experiments the α-rays hardly come into play, because these rays
are practically suppressed by very thin absorbent screens. Those
which penetrate are, on the one hand, the more or less scattered β-
rays; on the other hand, γ-rays, which seem analogous to Röntgen
rays.

The following are the results of some of my experiments on
this subject: —

The radium is enclosed in a glass vessel. The rays, that emerge
from the vessel, pass through 30 c.m. of air, and are received on
a series of glass plates, each of thickness 1.3 m.m.; the first plate
transmits 49 per cent of the radiation it receives, the second transmits 84 per cent of the radiation it receives, the third transmits
85 per cent of the radiation it receives.

In another series of experiments the radium was enclosed in a
glass vessel placed 10 c.m. from the condenser that received the
rays. A series of identical lead screens were placed on the vessel,
each one having a thickness of 0.115 m.m.

The ratio of the radiation transmitted to the radiation received
is given for each of the successive screens by the series of the
following numbers: —

---

[30]Meyer and von Schweidler, *Physik. Zeitschrift*, Vol. I, p. 209.

0.40   0.60   0.72   0.79   0.89   0.92   0.94   0.94   0.97

For a series of four screens of lead, each of which was 1.5 m.m. thick, the ratio of the radiation transmitted to the radiation received was given for the successive screens by the following numbers: —

0.09   0.78   0.84   0.82

From these experiments it follows that when the thickness of the lead traversed increases from 0.1 m.m. to 6 m.m., the penetrating power of the radiation increases.

I have observed that, under the experimental conditions indicated, a lead screen 1.8 c.m. thick transmits 2 per cent of the radiation it receives; a screen of lead 5.3 c.m. thick transmits 0.4 per cent of the radiation it receives. I also found that the radiation transmitted by a lead of thickness equal to 1.5 m.m. includes a high proportion of deviating rays (cathode type). The latter are therefore capable of crossing not only great distances in the air, but also considerable thicknesses of very absorbent solids, such as lead.

When we study with the device of Fig. 2 the absorption exerted by an aluminum screen of 0.01 m.m. thickness on the whole of the radium radiation, the screen being always placed at the same distance from the radiant substance, and the condenser being placed at a variable distance AD, the results obtained are the sum of those due to the three groups of the radiation. If observed from a long distance, the penetrating rays dominate and the absorption is low; if observed from a short distance, α-rays dominate and the absorption is lower the closer we get to the substance; for an intermediate distance, absorption goes through a maximum and penetration through a minimum.

Distance AD	7.1	6.5	6.0	5.1	3.4
Percentage of rays transmitted by aluminium	91	82	58	41	48

However, certain absorption-related experiments show a certain analogy between the α-rays and the β-rays. This is how M. Becquerel discovered that the absorbent action of a solid screen on the β-rays increases with the distance of the screen from the source, so that, if the rays are subjected to a magnetic field, as in Fig. 4, a screen placed in contact with the radiant source allows a larger portion of the magnetic spectrum to be in evidence than does the same screen placed upon the photographic plate. This variation in the absorbing effect of the screen with the distance of this screen from the source is analogous to what takes place for α-rays; this has been verified by MM. Meyer and von Schweidler, who operated by means of the fluoroscopic method; M. Curie and I observed the same fact when using the electrical method. The conditions for the production of this phenomenon have not yet been studied. However, when the radium is enclosed in a glass tube and placed at a distance from the condenser, which is itself enclosed in a thin aluminium box, it becomes irrelevant whether the screen is placed against the source or against the condenser; the current obtained is the same in both cases.

The investigation of the α-rays led me to consider that these rays behave like projectiles having a certain initial velocity, and which lose their force when crossing obstacles[31]. These rays, moreover, travel by rectilinear propagation, as has been shown by M. Becquerel in the following experiment: —

Polonium emitting rays was placed in a very narrow straight cavity hollowed in a sheet of cardboard. We thus had a linear

---

[31]Mme Curie, *Comptes rendus*, 8 January 1900.

source of rays. A copper wire, 1.5 m.m. in diameter, was placed parallel in front of the source at a distance of 4.9 m.m. A photographic plate was placed parallel, beyond, at a distance of 8.65 m.m. After an exposure of ten minutes, the geometric shadow of the wire was perfectly reproduced, with the dimensions provided and a narrow penumbra on each side corresponding to the size of the source. The same experiment succeeded equally well when a double sheet of beaten aluminium was placed against the wire, through which the rays were forced to pass.

So these are indeed rays capable of giving perfect geometric shadows. The experiment with the aluminium shows that these rays are not scattered through the screen, and that this screen does not emit, at least in significant quantities, secondary rays similar to the secondary rays of the Röntgen rays.

The α-rays are those which seem active in the very beautiful experiment carried out in the spinthariscope of M. Crookes[32]. This device essentially consists of a grain of radium salt held at the end of a metal wire in front of a phosphorescent zinc sulfide screen. The grain of radium is at a very small distance from the screen (0.5 m.m., for example), and one looks with a magnifying glass the face of the screen turned towards the radium. Under these conditions the eye sees on the screen a real rain of luminous points which appear and disappear continuously. The screen looks like a starry sky. The bright spots are closer together in the areas of the screen near radium, and in the immediate vicinity of the radium the glow appears continuous. The phenomenon does not seem to be affected by the air currents; it occurs in a vacuum; even a very thin screen placed between the radium and the phosphorescent screen suppresses the phenomenon; so it

---

[32] *Chem. News,* 3 April 1903

seems that the phenomenon is due to the action of the most absorbable α-rays of radium.

One can imagine that the appearance of one of the bright spots on the phosphorescent screen is caused by the impact of an isolated projectile. In this view, we would be dealing, for the first time, with a phenomenon making it possible to distinguish the individual action of a particle whose dimensions are of the same order of magnitude as those of an atom.

The appearance of bright spots is the same as that of stars or highly-lit ultra-microscopic objects which do not produce sharp images on the retina, but diffraction spots; and this is in good agreement with the concept that each extremely small point of light is produced by the shock of a single atom.

The undeflectable penetrating γ-rays seem to be of a completely different nature and seem similar to the Röntgen rays. There is nothing to prove, moreover, that little penetrating rays of the same nature cannot exist in the radiation of radium, because they could be masked by corpuscular radiation.

We have just seen how complex the radiation of radioactive bodies is. The difficulties of its study are increased by this circumstance, which it is necessary to investigate if this radiation experiences only a selective absorption by matter, or also a more or less deep transformation.

Little is known yet about this issue. However, if we assume that the radiation of radium includes both cathode rays and Röntgen rays, we can expect that this radiation will undergo transformations when passing through screens. We know, in fact, first that the cathode rays which exit from the Crookes tube through an aluminum window (Lenard's experiment) are strongly diffused by aluminum, and that, moreover, crossing the screen leads to a decrease in the speed of the rays; this is how cathode rays with a speed equal to 1.4×10^{10} centimeters lose 10 per cent of their speed when passing through 0.01 m.m. of aluminum[33]; second, the cathode rays, by striking an obstacle, give rise to the production of Röntgen rays; third, the Röntgen rays, by striking a solid obstacle, give rise to a production of secondary rays, which are partly cathode rays[34].

We can therefore, by analogy, predict the existence of all the above phenomena for the rays of radioactive substances.

By studying the transmission of polonium rays through an aluminum screen, M. Becquerel observed neither the production of secondary rays nor their transformation into cathodic rays[35].

I sought to demonstrate a transformation of the rays of polonium by using the method of inversion of the screens. Two superposed screens, E_1 and E_2 , being crossed by the rays, the order in which they are traversed should be immaterial if the passage through the screens does not transform the rays; if, on the contrary, each screen transforms the rays during transmission, the order of the screens is of moment. If, for example, the rays are transformed into more absorbable rays in passing through lead, and no such effect is produced by aluminium, then the system lead-aluminium will be more opaque than the system aluminium-lead; this takes place with Röntgen rays.

My experiments show that this phenomenon occurs with the rays of polonium. The apparatus used was that of Fig. 8. The polonium was placed in the box, $CCCC$, and the absorbent screens, necessarily very thin, were placed upon the metallic sheet T.

The results obtained prove that the radiation is modified in crossing a solid screen. This conclusion is in agreement with

---

[33]Des Coudres, *Physik. Zeitschrift*, November 1902
[34]Sagnac, *Thèse de doctorat.* — Curie and Sagnac, *Comptes rendus*, April 1900.
[35]Becquerel, *Rapports au Congrès de Physique*, 1900.

Screens employed.	Thickness. m.m.	Current observed.
Aluminium Brass	0.01 0.005	17.9
Brass Aluminium	0.005 0.01	6.7
Aluminium Tin	0.01 0.005	150
Tin Aluminium	0.005 0.01	125
Tin Brass	0.005 0.005	13.9
Brass Tin	0.005 0.005	4.4

the experiments in which, of two identical superposed metallic screens, the first is less absorbent than the second. From this it is probable that the transforming action of a screen increases with the distance of the screen from the source. This fact has not been verified, and the nature of the transformation has not been studied in detail.

I repeated the same experiments with a very active salt of radium; the result was negative. I have only observed insignificant variations in the intensity of the radiation transmitted with interchange of the order of the screens. The screen systems tested were as follows: —

Aluminium, thickness 0.55 m.m.
Aluminium, thickness 0.55 m.m.
Aluminium, thickness 0.55 m.m.
Aluminium, thickness 1.07 m.m.
Aluminium, thickness 0.55 m.m.

Aluminium, thickness	1.07	m.m.
Aluminium, thickness	0.15	m.m.
Aluminium, thickness	0.15	m.m.
Aluminium, thickness	0.15	m.m.
Platinum, thickness	0.01	m.m.
Lead, thickness	0.1	m.m.
Tin, thickness	0.005	m.m.
Copper, thickness	0.05	m.m.
Brass, thickness	0.005	m.m.
Brass, thickness	0.005	m.m.
Platinum, thickness	0.01	m.m.
Zinc, thickness	0.05	m.m.
Lead, thickness	0.1	m.m.

The system lead-aluminium was slightly more opaque than the system aluminium-lead, but the difference was not large.

I could not thus highlight a notable transformation of the rays of radium. However, in various radiographic experiments, M. Becquerel observed very intense effects due to scattered or secondary rays, emitted by solid screens which received radium rays. Lead seemed to be the most active substance in this respect.

Ionising Action of Radium Rays on Insulating Liquids.

M. Curie has shown that radium rays and Röntgen rays act on liquid dielectrics as on air, by imparting to them a certain electrical conductivity[36]. The experiment was arranged in the following way (Fig. 9).

The liquid to be tested is placed in a metal vessel, *CDEF*, into which a thin copper tube, *AB*, is immersed; these two metal parts

---

[36]P. Curie, *Comptes rendus de l'Académie des Sciences*, 17 February 1902.

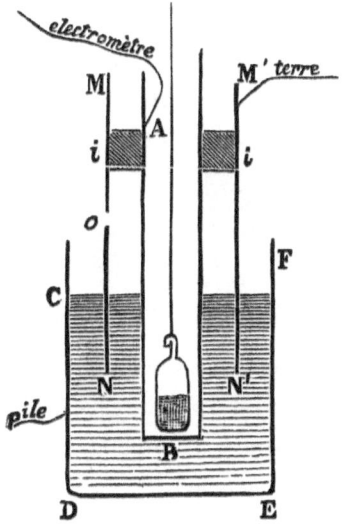

Figure 9.

serve as electrodes. The outer vessel is maintained at a known potential by means of a battery of small accumulators, one pole of which is connected to earth. The tube, AB, is connected to the electrometer. When a current flows through the liquid, the electrometer is kept at zero using a piezoelectric quartz, which gives the strength of the current. The copper tube, $MNM'N'$, connected to the ground, serves as a guard tube to prevent the flow of current through the air. A bulb containing the radium-barium salt can be placed at the bottom of the tube, AB; the rays act on the liquid after passing through the glass of the bulb and the walls of the metal tube. The radium can still be allowed to act by placing the bulb below the side, DE.

Working with Röntgen rays, we make these rays arrive through the wall DE.

The increase of conductivity by the action of radium rays or the Röntgen rays seems to occur for all liquid dielectrics; but in order to observe this increase, it is necessary that the inherent conductivity of the liquid must be low enough to not mask the effect of the rays.

By operating with radium and Röntgen rays, M. Curie obtained effects of the same order of magnitude.

When we investigate with the same device the conductivity of air or of another gas under the action of the Becquerel rays, we find that the intensity of the current obtained is proportional to the difference of potential between the electrodes, as long as the latter does not exceed a few volts; but at higher voltages, the intensity of the current increases less and less rapidly, and the saturation current is practically attained for a voltage of 100 volts.

The liquids examined with the same device and the same very radioactive product behave differently; the intensity of the current is proportional to the tension when the latter varies between 0 and 450 volts, and when the distance between the electrodes does not exceed 6 m.m. We can then consider the conductivity caused in various liquids by the radiation of a radium salt acting under the same conditions

The numbers in the following table multiplied by 10^{-11} give the conductivity in mhos (inverse of ohm) per c.c.: —

Carbon bisulphide	20
Petroleum ether	15
Amylene	14
Benzine	4
Liquid air	1.3
Vaseline oil	1.6

We can, however, assume that liquids and gases behave similarly, but that, in the case of liquids, the current remains proportional to the voltage up to a much higher limit than for gases. One could, by analogy with what is observed for gases, seek to lower the limit of proportionality by using a much weaker radiation. Experiments have verified this prediction. The radioactive product employed was 150 times less active than that used for the previous experiments. For tensions of 50, 100, 200, 400 volts, the current intensities were represented respectively by the numbers 109, 185, 255, 335. The proportionality was no longer maintained, but the current showed great variation when the difference of potential was doubled.

Some of the liquids examined are nearly perfect insulators, when kept at a constant temperature, and when screened from the action of rays. These are liquid air, petroleum ether, vaseline oil, and amylene. It is then very easy to study the effect of rays. Vaseline oil is much less sensitive to the action of rays than petroleum ether. This fact may have some relation to the difference in volatility which exists between these two hydrocarbons. Liquid air, which has boiled for some time in the experimental vessel, is more sensitive to the action of the rays than that which has just been poured into it; the conductivity produced by the rays is one-fourth as great again in the former case. M. Curie has investigated the action of the rays upon amylene and upon petroleum ether at temperatures of $+10°$ and $-17°$. The conductivity due to the radiation diminishes by one-tenth of its value only, in passing from $10°$ to $-17°$.

In experiments where the temperature of the liquid is varied, we can either keep the radium at room temperature or bring it to the same temperature as the liquid; the same result is obtained in both cases. This is because the radiation of radium does not vary with the temperature, and remains unaltered even at the temperature of liquid air. This fact has been verified directly by measurements.

Various Effects and Applications of the Ionising Action of Rays Emitted by Radioactive Substances.

The rays of the new radioactive substances have a strong ionising action on air. The action of radium can easily cause *the condensation of supersaturated water vapour*, just as it does through the action of cathode rays and Röntgen rays.

Under the influence of the rays emitted by the new radioactive substances, the *distance of discharge between two metallic conductors for a given potential difference is increased*; in other words, the passage of the spark is facilitated by the action of the rays. This phenomenon is due to the action of the most penetrating rays. If, in fact, the radium is surrounded by a lead envelope of 2 c.m., the action of radium on the spark is not considerably weakened, while the radiation which passes through is only a very small fraction of the total radiation.

In causing conductivity by the action of radioactive substances, the air in the neighbourhood of two metallic conductors, one of which is connected to earth and the other to a well-insulated electrometer, we see the electrometer to be permanently deflected, which makes it possible to measure the electro-motive force of the battery formed by the air and the two metals (electromotive force of contact of the two metals, when they are separated by the air). This method of measurement was employed by Lord Kelvin and his students, the radiating substance being uranium[37]; a similar method had previously been employed by M. Perrin, who used the ionising action of the Röntgen rays[38].

---

[37]Lord Kelwin, Beattie and Smolan, *Nature*, 1897.
[38]Perrin, *Thèse de doctorat.*

Radioactive substances can be used in the study of atmospheric electricity[39]. The active substance is enclosed in a small thin aluminium box attach at the end of a metal rod connected with the electrometer. The air is made to conduct in the vicinity of the end of the rod, and this takes the potential of the surrounding air. Radium thus advantageously replaces the flames or the water-flow devices of Lord Kelvin, generally used hitherto in the investigation of atmospheric electricity.

Fluorescent and Luminous Effects.

The rays emitted by the new radioactive substances cause the fluorescence of certain bodies. M. Curie and I first discovered this phenomenon by making the act through a layer of aluminium foil placed on a layer of barium platinocyanide. The same experiment succeeds even more easily with sufficiently active barium containing radium. When the substance is strongly radioactive the fluorescence produced is very beautiful.

A large number of bodies are capable of becoming phosphorescent or fluorescent by the action of the Becquerel rays. M. Becquerel studied the effect on uranium salts, diamonds, blende, &c. M. Bary has demonstrated that the salts of alkalis and alkaline earth metals, which are all fluorescent under the action of light rays and Röntgen rays, are also fluorescent under the action of radium rays[40]. We can also observe the fluorescence of paper, cotton, glass, &c., in the vicinity of radium. Among the different types of glass, Thuringian glass is particularly bright. Metals do not seem to become luminous.

---

[39]Paulsen, *Rapports au Congrès de Physique*, 1900. — Witkowski, *Bulletin de l'Académie des Sciences de Cracovie*, January 1902.

[40]Bary, *Comptes rendus*, Vol. CXXX, 1900, p. 776.

Barium platinocyanide is best suited for studying the radiation of the radioactive substances by the fluoroscopic method. The effect of the radium rays may be followed at distances greater than 2 m. Phosphorescent zinc sulphide is made extremely bright, but this body has the disadvantage of preserving its luminosity for some time after the action of the rays has ceased.

The fluorescence produced by radium can be observed when the fluorescent screen is separated from the radium by absorbent screens. We were able to observe the illumination of a screen of barium platinocyanide through the human body. However, the action is incomparably greater when the screen is placed right against the radium and it is not separated from it by any solid screen. All the groups of rays seem capable of producing fluorescence.

To observe the action of polonium, it is necessary to put the substance very close to the fluorescent screen without the interposition of a solid screen, or at least with the interposition of a very thin screen only.

The luminosity of fluorescent substances exposed to the action of radioactive substances decreases over time. At the same time the fluorescent substance undergoes a transformation. Here are few examples: —

Radium rays transform barium platinocyanide into a less luminous brown variety (an action similar to that produced by Röntgen rays, and described by M. Villard).

Uranium sulphate and potassium sulphate are similarly altered, turning into a yellow substance. The transformed barium platinocyanide is partially regenerated by the action of light. If the radium is placed beneath a layer of barium platinocyanide spread on paper, the platinocyanide becomes bright; if the system is kept in the dark, the platinocyanide will deteriorate, and its luminosity will drop considerably. But if the whole is exposed to light, the platinocyanide is partially regenerated, and if the whole is put back in darkness the brightness re-appears fairly strong. By means of a fluorescent body and a radioactive body, we have therefore obtained a system which acts as a phosphorescent body capable of long duration of phosphorescence.

Glass which is made fluorescent by the action of radium turns brown or purple. At the same time, it becomes less fluorescent. If this altered glass is heated, it discolors and, at the same time as discoloration occurs, the glass emits light. After that, the glass resumed the property of being fluorescent to the same degree as before the transformation.

Zinc sulphide, which has been exposed for a sufficient time to the action of radium, gradually becomes exhausted and loses the ability to be phosphorescent, either under the action of radium or that of light.

The diamond becomes phosphorescent by the action of radium, and can thus be distinguished from rhinestone imitations, whose brightness is very low.

All the barium-radium compounds *are spontaneously luminous*[41]. The dry anhydrous halogen salts emit a particularly intense light. This brightness cannot be seen in broad daylight, but it is easily seen in the semi-darkness or in a room lit by gas-light. The light emitted can be strong enough so one can read in the dark. The light emitted emanates from the entire mass of the product, whilst for an ordinary phosphorescent body, the light emanates especially from the portion of the surface which has been illuminated. Radium products lose much of their luminosity in humid air, but they regain it on drying (Giesel). The

---

[41]Curie, *Soc. de Physique*, 3 March 1899 — Giesel, *Wied. Ann.*, Vol. LXIX, p. 91.

brightness seems to be preserved. After several years no sensible change seems to have occurred in the brightness of weakly active products, kept in the dark in sealed tubes. In the case of very active and very luminous radium-barium chloride, the light changes colour after several months; it becomes more violet and loses in intensity; at the same time the product undergoes transformations; on re-dissolving the salt in water and drying it afresh, the original luminosity is restored.

Solutions of barium-radium salt, which contain a large proportion of radium, are equally luminous; this fact may be observed by placing the solution in a platinum capsule, which not being itself luminous permits of the faint luminosity of the solution being seen.

When a solution of a barium-radium salt contains crystals deposited in it, these crystals are luminous at the bottom of the solution, and much more so than the solution itself, so that they alone appear luminous.

M. Giesel has made a preparation of barium-radium platinocyanide. When this salt has just crystallised, it looks like ordinary barium platinocyanide and it is very bright. But gradually the salt becomes spontaneously coloured, taking a brown tint, at the same time the crystals becoming dichroic. In this state the salt is much less luminous, although its radioactivity has increased[42]. The radium platinocyanide, prepared by M. Giesel, changes even more rapidly.

Radium compounds are the first example of spontaneously luminous substances.

---

[42]Giesel, *Wied. Ann.*, Vol. LXIX, p. 91.

Heat Generation by Radium Salts.

MM. Curie and Laborde have recently discovered that *the salts of radium are the source of a spontaneous and continuous heat generation*[43]. This heat release has the effect of keeping the salts of radium at a temperature higher than room temperature; the excess also depends on the thermal insulation of the substance. This excess temperature can be demonstrated by a rough experiment using two ordinary mercury thermometers. Two thermal vacuum insulating vessels, identical to each other, are used. A glass bulb is placed in one of the vessels containing 7 d.g. of pure radium bromide; in the second vessel is placed another similar glass bulb which contains some inactive substance, for example barium chloride. The temperature of each enclosure is indicated by a thermometer, placed in the immediate vicinity of the bulb. The opening of the insulators is closed with cotton. When the temperature equilibrium is established, the thermometer, which is in the same vessel as the radium, constantly indicates a temperature higher than that indicated by the other thermometer; the excess temperature observed was 3°.

The amount of heat given off by radium can be estimated using a Bunsen ice calorimeter. By placing a glass bulb containing the radium salt in this calorimeter, there is a continuous supply of heat which stops as soon as the radium is removed. Measuring with a long-prepared radium salt indicates that each gram of radium releases about 80 small calories during each hour. Radium therefore give off an amount of heat sufficient to melt its weight of ice in one hour, and one grm.-atom (225 grm.) of radium releases 18,000 cal., a quantity of heat comparable to that produced by the combustion of 1 grm.-atom (1 grm.) of hydrogen. Such

---

[43]Curie and Laborde, *Comptes rendus*, 16 March 1903.

a considerable heat flow cannot be explained by any ordinary chemical reaction, and all the more so since the state of radium seems to remain unaffected for years. The evolution of heat might be attributed to a slow transformation of the radium atom. If this were the case, we should be led to conclude that the quantities of energy generated during the formation and transformation of the atoms are considerable, and that they exceed all that is so far known.

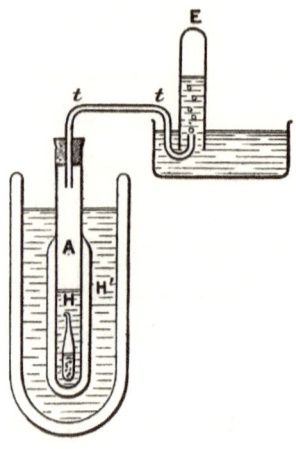

Figure 10.

We can also assess the heat given off by radium at various temperatures by using it to boil a liquefied gas and by measuring the volume of the gas which is released. We can do this experiment with methyl chloride (at −21°). The experiment was carried out by MM. Dewar and Curie with liquid oxygen (at −180°) and with liquid hydrogen (at −252°). This latter body is particularly suitable for carrying out the experiment. A test tube *A*, surrounded by a vacuum thermal insulator, contains liquid hydrogen *H* (fig. 10); it is equipped with a release tube *t* which allows the gas to be collected in a graduated cylinder *E* filled with water. The test tube A and its insulator immerse in a bath of liquid hydrogen H'. Under these conditions no gas evolution occurs in the test tube A. When one introduces, in the hydrogen, liquid contained in this test tube, a bulb which contains 7 d.g. of radium bromide, there is a continuous release of gas, and 73 cm^3 of gas are collected per minute.

A solid radium salt which has just been prepared gives off a relatively small amount of heat; but this heat flow increases continuously and tends towards a determined value which is not yet fully reached after a month. When a radium salt is dissolved in water and the solution is sealed in a sealed tube, the amount of heat given off by the solution is initially small; it then increases and tends to become constant after one month; the heat flow is then the same as that due to the same solid salt.

When you measure the heat given off by a radium salt contained in a glass bulb using the Bunsen calorimeter, certain penetrating rays of radium pass through the bulb and the calorimeter without being absorbed there. To see if these rays take up an appreciable amount of energy, we can redo a measurement by surrounding the bulb with a sheet of lead 2 m.m. thick; it is found that, under these conditions, the release of heat is increased by about 4 per cent of its value; the energy emitted by radium in the form of penetrating rays is therefore by no means negligible.

Chemical Effects produced by the New Radioactive Substances.

Colourations. — The radiations emitted by strongly radioactive substances are capable of causing certain transformations, certain chemical reactions. The rays emitted by radium products exercise colouring actions upon glass and porcelain[44].

The colouration of glass, generally brown or purple, is very intense; it is produced in the very mass of the glass, and it persists after the radium is removed. All the glasses are coloured after a longer or a shorter interval of time, and the presence of lead is not necessary. This fact should be compared with that, recently observed, of the coloring of the glasses of the vacuum tubes, after having been in use for a long time for the production of Röntgen rays.

M. Giesel has shown that the crystallised halogen salts of the alkali metals (rock salt, sylvite) become coloured under the influence of radium, as under the action of cathode rays. M. Giesel points out that similar colourations are obtained by letting the salts of the alkalis be exposed to sodium vapour[45].

I investigated the colouration of a collection of glasses of known composition, kindly lent me for the occasion by M. Le Chatelier. I observed no great variety in the colouration. It is generally purple, yellow, brown, or grey. It seems to be linked to the presence of alkali metals.

With the pure crystallised alkali salts we obtain more varied and more vivid colours; the salt, originally white, becomes blue, green, yellow, brown, &c.

M. Becquerel has discovered that white phosphorus is transformed into red phosphorus by the action of radium.

Paper is altered and coloured by the action of radium. It becomes fragile, crumbles and finally resembles a colander perforated with holes.

Under certain circumstances, ozone is produced in the vicinity of very active compounds. The rays emerging from a sealed bulb,

---

[44]M. and M[me] Curie, *Comptes rendus*, Vol. CXXLX, November 1899, p. 828.
[45]Giesel, *Soc. de Phys. allemande*, January 1900.

containing radium, do not produce ozone in the air they pass through. On the contrary, a strong odour of ozone is released when the bulb is opened. Generally speaking, ozone is produced in the air when the latter is in direct contact with the radium. Communication through even an extremely narrow conduit is sufficient; ozone production appears to be linked to the propagation of induced radioactivity, of which will be discussed below.

Radium compounds appear to change over time, presumably under the action of their own radiation. We have seen above that crystals of barium-radium chloride, which are colourless at the time of deposition, gradually take on a colour sometimes yellow or orange, sometimes pink; this colouration disappears in solution. Barium-radium chloride generates oxygen compounds of chlorine; the bromide those of bromine. These slow transformations generally take place some time after the preparation of the solid product, which, at the same time, changes its appearance and colour, taking on a yellow or purple tint. The light emitted also becomes more purple.

Pure radium salts seem to undergo the same transformations as those containing barium. However, crystals of the chloride, deposited in acid solution, do not colour appreciably after some time has elapsed, whereas crystals of barium-radium chloride, rich in radium, become deeply coloured.

Gas Emission in the Presence of Radium Salts. — A solution of radium bromide gives off gases continuously[46]. These gases are mainly hydrogen and oxygen, and the composition of the mixture is close to that of water; we can assume that water decomposes in the presence of the radium salt.

Solid radium salts (chloride, bromide) also give rise to a con-

---

[46]Giesel, *Ber.*, 1903, p. 347. — Ramsay and Soddy, *Phys. Zeitschr.*, 15 September 1903.

tinuous emission of gas. These gases fill the pores with solid salt and are released quite abundantly when the salt is dissolved. We find in the gas mixture hydrogen, oxygen, carbonic acid, helium; the gas spectrum also presents some unknown lines[47].

Two accidents that occurred in M. Curie's experiments can be attributed to gas releases. A sealed thin glass ampoule, almost completely filled with dry, solid radium bromide, exploded two months after closing it due to little heating; the explosion was likely due to the pressure of the interior gas. In another experiment, a bulb containing radium chloride, prepared since a long time, communicated with a fairly large tank in which a very perfect vacuum was maintained. The bulb having been subjected to a fairly rapid heating up to 300°, the salt exploded; the bulb was broken, and the salt was thrown to a large distance; there could be no noticeable pressure in the bulb at the time of the explosion. The device had also been subjected to a heating test under the same conditions in the absence of the radium salt, and no accident had occurred.

These experiments show that it is dangerous to heat radium salt that has been prepared since a long time and that it is also dangerous to keep radium in a sealed tube for a long time.

Production of Thermoluminescence. — Some bodies, such as fluorite, become luminous when heated; they are thermo-luminescent. Their luminosity disappears after some time, but the faculty of becoming luminous again through heat is restored to these bodies by the action of a spark and also by the action of radium. Radium can thus restore to these bodies their thermo-luminescent property[48]. Fluorite when heated undergoes a transformation, which is accompanied by the emission of light. If the

---

[47]Ramsay and Soddy, *loc. cit.*.
[48]Becquerel, *Rapports au Congrès de Physique*, 1900.

fluorite is afterwards subjected to the action of radium, an inverse change occurs, which is also accompanied by an emission of light.

An absolutely analogous phenomenon occurs when glass is exposed to radium rays. Here too a transformation occurs in the glass, while it is luminous under the action of radium rays; this transformation is highlighted by the colouration which appears and gradually increases. If the glass is afterwards heated, the inverse change takes place, the colour disappears, and this phenomenon is accompanied by production of light. It appears very probable that we have here a change of a chemical nature, and the production of light is associated with this change. This phenomenon may be general. It might be that the production of fluorescence by the action of radium and the luminosity of radium compounds is of necessity associated with some chemical or physical change in the substance emitting the light.

Radiographs. — The radiographic action of the new radioactive substances is very intense. However, the way of operating must be very different with polonium and radium. Polonium acts only at a very short distance, and its action is considerably weakened by solid screens; it is easily suppressed by means of a thin screen (1 m.m. of glass).

Radium acts at considerably greater distances. The action of radium rays may be observed at more than 2 m. distance in air, even when the active product is enclosed in a glass vessel. The rays acting under these conditions belong to the β- and γ-groups. Thanks to the differences in transparency of different materials to the rays, one can, as with Röntgen rays, obtain radiographs of different objects. Metals are, in general, opaque, with the exception of aluminium, which is very transparent. There is no noticeable difference in transparency between flesh and bone. We can operate at great distance and with a source of very small di-

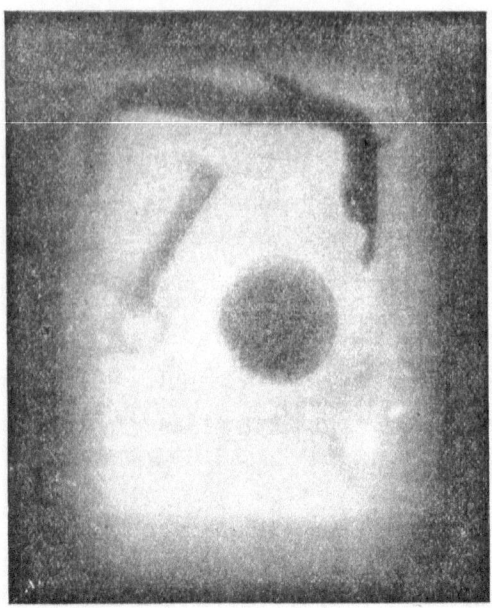

Figure 11.: Radiography obtained with the rays of radium. (This picture shows
Marie Curie's purse, exposed to radium rays – Translator's note)

mensions; and then we have very fine radiographs. The beauty of
the radiograph is enhanced by deflecting to one side the β-rays, by
means of a magnetic field, and to use only the γ-rays. The β-rays,
in in passing through the object to be radiographed, undergo
a certain amount of diffusion, and thus cause a slight blurring.
By removing them, a longer time of exposure is necessary, but
better results are obtained. The radiograph of an object, such as
a purse, requires one day with a radiating source composed of
several centigrms. of a radium salt, enclosed in a glass vessel, and
placed at a distance of 1 m. from the sensitive plate, in front of
which the object is placed. If the source is at a distance of 20 c.m.
from the plate, the same result is obtained in one hour. In the
immediate vicinity of the source of radiation, a sensitive plate is
instantaneously impressed.

Physiological Effects.

Radium rays exert an action on the epidermis. This action has been observed by M. Walkhoff and confirmed by M. Giesel, then by MM. Becquerel and Curie[49].

If a celluloid or a thin indiarubber capsule containing a very active salt of radium is placed on the skin and left there for some time, a redness is produced upon the skin, either immediately or at the end of some time, which is longer in proportion as the action is weaker; this red spot appears in the place which has been exposed to the action; local skin alteration manifests and progresses like a burn. In certain cases a blister forms. If the exposure has been prolonged, an ulceration is produced which takes a very long time to heal. In one experiment, M. Curie had a relatively weak radioactive product to act on his arm for ten hours. The redness appeared immediately, and a wound formed later which took four months to heal. The epidermis was destroyed locally, and could only be restored to the healthy state slowly and painfully, leaving a very marked scar. A radium burn with half-an-hour's exposure appeared after fifteen days, formed a blister and healed in fifteen days. Another burn, made with an exposure of only eight minutes, caused a red spot which appeared after only two months and its effect was insignificant.

The action of radium upon the skin can occur through metals, but it is weakened. To guarantee action, you must avoid keeping radium on you for a long time other than wrapped in lead foil.

The action of radium upon the skin has been investigated by Dr. Daulos, at the Hospital of S. Louis, as a method of treat-

---

[49]Walkhoff, *Phot. Rundschau*, October 1900. — Giesel, *Berichte d. deutsch. chem. Gesell.*, Vol. XXIII. — Becquerel and Curie, *Comptes rendus*, Vol. CXXXII, p. 1289.

ing certain skin diseases, a process comparable to the treatment with the Röntgen rays or the ultra-violet rays. From this point of view, radium gives encouraging results; the epidermis partially destroyed by the action of radium re-forms in a healthy state. The action of radium is more penetrating than that of light, and its use is easier than that of light or of Röntgen rays. The study of the conditions of application is of necessity rather lengthy, because the effect of the application does not appear at once.

M. Giesel has observed the action of radium on plant leaves. The leaves subjected to the action turn yellow and crumble.

M. Giesel also discovered the action of radium rays on the eye[50]. When you place a radioactive substance in the dark, near your closed eyelid or temple, a sensation of light fills the eye. This phenomenon has been studied by MM. Himstedt and Nagel[51]. These physicists have demonstrated that the centre of the eye becomes fluorescent by the action of radium, and this is what explains the sensation of light experienced. Blind people whose retina is intact are sensitive to the action of radium, while those whose retina is damaged do not experience any sensation of luminosity due to the rays.

Radium rays prevent or hinder the development of microbial colonies , but this action is not very intense[52].

M. Danysz has recently demonstrated that the rays of radium act energetically on the marrow and on the brain. After one hour's exposure paralysis occurs in animals tested, and these usually die within a few days[53].

---

[50]Giesel, *Naturforscherrersammlung*, München, 1899.
[51]Himstedt and Nagel, *Ann. der Physik*, Vol. IV, 1901.
[52]Aschkinass and Caspari, *Ann. der Physik*, Vol. VI, 1901, p. 570.
[53]Danysz, *Comptes rendus*, 16 February 1903.

Influence of Temperature on Radiation.

We still have little information on how the emission of radioactive bodies varies with temperature. We do know, however, that radiation persists at low temperatures. M. Curie placed a glass tube containing barium-radium chloride in liquid air[54]. The luminosity of the radioactive body persisted under these conditions. By the time the tube is removed from the cold bath, it even appears brighter than at room temperature. At the temperature of liquid air radium continues to excite the fluorescence in the sulphates of uranium and potassium. M. Curie verified by electrical measurements that the radiation, measured at a certain distance from the source, has the same intensity whether the radium is at room temperature or at that of liquid air. In these experiments the radium was placed at the bottom of a tube closed at one end. The rays emerged from the open end of the tube, passed through a certain space in the air, and were collected in a condenser. The action of the rays on the air of the condenser was measured, either by leaving the tube in the air, or by surrounding it with liquid air up to a certain height. The result obtained was the same in both cases.

When radium is brought to a high temperature its radioactivity remains. Barium-radium chloride after being melted (around 800°) is radioactive and luminous. However, prolonged heating at a high temperature has the effect of temporarily lowering the radioactivity of the product. The decrease is very significant, it can constitute 15 per cent of the total radiation. The proportional decrease is less great for the absorbable rays than for the penetrating rays, which are to some extent suppressed by heating. Over time, the radiation of the product regains the intensity

---

[54]Curie, *Société de Physique*, 2 March 1900.

and composition that it possessed before heating; this result is achieved after about two months from heating.

Chapter IV. Induced Radioactivity

Communication of Radioactivity to Substances Initially Inactive.

During our research on radioactive substances M. Curie and I have noticed that any substance which remains for some time in the vicinity of a radium salt becomes itself radioactive[1]. In our first publication on this subject, we intended to prove that the radioactivity thus acquired by originally inactive substances is not due to a transport of radioactive dust which would have come to settle on the surface of these substances. This fact, proven beyond doubt, is clearly demonstrated by all the experiments which will be described here, and in particular by the laws according to which the radioactivity excited in naturally inactive substances disappears when these substances are removed from the action of radium.

We have given the name of *induced radioactivity* to the new phenomenon thus discovered.

In the same publication, we have indicated the essential characteristics of induced radioactivity. We excited screens of different substances by placing them in the neighbourhood of solid radium salts, and we investigated the radioactivity of these screens by the electrical method. We observed the following facts: —

1. The activity of a screen exposed to the action of radium

---

[1] M. and Mme Curie, *Comptes rendus*, 6 November 1899.

increases with the exposure time by approaching a definite limit, according to an asymptotic law.

2. The activity of a screen which was activated by the action of radium, and which is afterwards withdrawn from this action, disappears in a few days. This induced activity tends to zero as a function of time, according to an asymptotic law.

3. All conditions being equal, the radioactivity induced by the same radiferous product upon different screens is independent of the nature of the screen. Glass, paper, metals, are activated with the same intensity.

4. The radioactivity induced in one screen by differing radium products has a limiting value which rises with the increased activity of the product.

Shortly afterwards, Mr. Rutherford published a research from which it follows that compounds of thorium are capable of producing the phenomenon of induced radioactivity[2]. Mr. Rutherford found for this phenomenon the same laws as those which have just been exposed, and he discovered besides this important fact, that the bodies charged with negative electricity are activated more energetically than the others. Mr. Rutherford also observed that air passed over thorium oxide retains significant conductivity for about ten minutes. Air in this condition communicates induced radioactivity to inactive substances, especially to those negatively charged. Mr. Rutherford interprets his experiments by assuming that thorium compounds, especially the oxide, emit a particular radioactive *emanation* capable of being

---

[2]Rutherford, *Phil. Mag.*, January and February 1900.

carried by air currents and charged with positive electricity. This emanation would be the cause of the induced radioactivity. M. Dorn has reproduced, with salts of barium containing radium, the experiments that Mr. Rutherford had made with thorium oxide[3].

M. Debierne has shown that actinium causes extremely intense induced activity in nearby bodies. As in the case of thorium, there is a considerable carriage of activity by air currents[4].

Induced radioactivity has various aspects, and irregular results are obtained when the activity of a substance in the neighbourhood of radium is excited in open air. MM. Curie and Debierne have noticed that the phenomenon is, on the contrary, very regular when operating in isolation; they therefore investigated induced activity in a closed enclosure[5].

Activity Induced in an Closed Enclosure.

The induced radioactivity is both more intense and more regular when operating in isolation. The active material is placed in a little glass ampoule, *a*, open at *o* (Fig. 12), in the middle of a closed enclosure. Various plates, *A*, *B*, *C*, *D*, *E*, placed in the enclosure become radioactive after one day of exposure. The activity is the same, whatever the nature of the plate, for equal dimensions (lead, copper, aluminium, glass, ebonite, wax, cardboard, paraffin). The activity of one side of one of the plates is greater, the greater the free space opposite this side.

If we repeat the previous experiment with the bulb, *a*, completely closed, we obtain no induced activity.

[3]Dorn, *Abh. Naturforsch. Gesell.*, Halle, June 1900.
[4]Debierne, *Comptes rendus*, 30 July 1900; 16 February 1903.
[5]Curie and Debierne, *Comptes rendus*, 4 March 1901.

The radiation of radium does not directly affect the production of induced radioactivity. For this reason, in the preceding experiment the plate D, screened from the radiation by a lead plate of thickness, PP, is made as active as B and E.

Radioactivity is transmitted by the air by degrees from the radiating body to the body to be activated. It can even be transmitted far away through very narrow capillary tubes.

The induced radioactivity is both more intense and more regular if the activating solid radium salt is replaced by its aqueous solution.

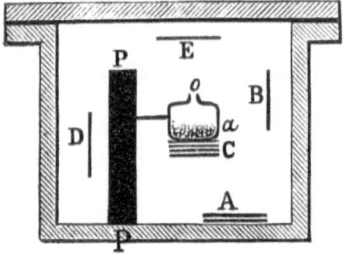

Figure 12.

Liquids are capable of acquiring induced radioactivity. For example, pure water can be made radioactive by placing it in a vase inside a closed enclosure that also contains a solution of radium salt. Certain substances become luminous when placed in an activating enclosure (phosphorescent and fluorescent bodies, glass, paper, cotton, water, saline solutions). The phosphorescent zinc sulphide is particularly bright under these conditions. The radioactivity of these luminous bodies is, however, the same as that of a piece of metal or other body which is excited under the same conditions without becoming luminous.

Whatever substance is activated in isolation, this substance acquires an activity which increases over time until it reaches a *limiting value*, always the same, when one operates with the same material and the same experimental arrangement.

The limit induced radioactivity is independent of the nature and pressure of the gas inside the activating enclosure (air, hydrogen, carbonic acid).

The limit induced radioactivity in the same enclosure depends only on the amount of radium present in the solution state, and seems to be proportional to it.

Role of Gases in the Phenomena of Induced Radioactivity.

Emanation. — The gases present in an enclosure which contains a solid salt or a solution of radium salt are radioactive. This radioactivity persists when the gas is sucked up with a tube and collected in a test-tube. The walls of the test-tube then become radioactive themselves, and the glass of the test-tube is luminous in the dark. The activity and the luminosity of the test-tube then disappear completely, but very slowly, and after a month we can still observe radioactivity.

Since the beginning of our research, M. Curie and I have extracted, by heating the pitchblende, a highly radioactive gas, but, as in the previous experiment, the activity of this gas at the end completely disappeared[6].

Thus, for thorium, radium, and actinium induced radioactivity is progressively propagated through the gases, from the active body to the walls of the enclosure which contains it, and the activating property is entrained with the gas itself, when we extract it from the enclosure.

---

[6]P. Curie and Mme Curie, *Rapports au Congrès de Physique*, 1900.

When the radioactivity of radium matter is measured by the electrical method using the apparatus of Fig. 1, the air between the plates also becomes radioactive; however, by sending an air current between the plates, there is no observable drop in the intensity of the current, which proves that the radioactivity spread in the space between the plates is of little importance in comparison with that of the radium itself in the solid state.

It is quite different in the case of thorium. The irregularities which I observed when determining the radioactivity of the thorium compounds stemmed from the fact that at this point I was working with a condenser in open air; thus, the slightest air current produces a considerable change in the intensity of the current, because the radioactivity spread in the space near the thorium is important compared to the radioactivity of the substance.

This effect is even more marked for actinium. A very active actinium compound appears to be much less active when we pass a stream of air over the substance.

The radioactive energy is therefore contained in gases in a special form. Mr. Rutherford assumed that radioactive bodies constantly give off a radioactive substance which he called emanation. It is this gas which would have the property of making radioactive the bodies which are in the space where it is spread. The bodies that emit the emanation are radium, thorium and actinium.

Deactivation of Activated Solid Bodies in Open Air.

A solid body, which has been activated by radium in an activating enclosed space for a sufficient time, and which has then been removed from the enclosure, deactivates in open air according to an exponential law, which is the same for all bodies represented by the following formula[7]: —

$$I = I_0 \left(a e^{-\frac{t}{\theta_1}} - (a-1) e^{-\frac{t}{\theta_2}} \right).$$

I_0 being the initial intensity of the radiation when the plate is removed from the enclosure; I, the intensity at time, t; a is a numerical coefficient, $a = 4.20$; θ_1 and θ_2 are time constants, $\theta_1 = 2420$ secs., $\theta_2 = 1860$ secs. After two or three hours this law becomes practically a simple exponential: the effect of the second exponential on the value of I is negligible. The deactivation law is therefore such that the intensity of radiation drops by half its value in twenty-eight minutes. This final law may be considered characteristic of the deactivation in the open air of solid bodies activated by radium.

The solid bodies activated by actinium lose their activity in the open air according to an exponential law similar to the previous one. However, deactivation is a little slower[8].

Solids activated by thorium, lose their activity much more slowly; the intensity of the radiation decreases by half in eleven hours[9].

Dissipation of Activity in a Confined Space. Velocity of Destruction of the Emanation.[10]

An enclosure, made active by radium and then removed from its influence, loses its activity according to a law much slower than that of deactivation in the open air. One can, for example,

---

[7]Curie and Danne, *Comptes rendus*, 9 February 1903.
[8]Debierne, *Comptes rendus*, 16 february 1903.
[9]Rutherford, *Phil. Mag.*, February 1900.
[10]P. Curie, *Comptes rendus*, 17 November 1902.

experiment with a glass tube, which is activated internally, by placing it for some time in contact with a solution of radium salt. The tube is then sealed with a lamp, and the intensity of radiation emitted by the walls of the tube is measured while the deactivation occurs.

The deactivation law is an exponential law. It is given with great accuracy by the formula

$$I = I_0 e^{-\frac{t}{\theta}}$$

I_0 = initial intensity of radiation. I = intensity of radiation at time, t. θ = a time constant, $\theta = 4.970 \times 10^5$ secs.

The intensity of the radiation decreases by half in four days.

This law of deactivation is absolutely invariable, whatever the conditions of the experiment (dimensions of the enclosure, nature of the walls, nature of the gas in the enclosure, duration of activation, &c.). The deactivation law remains the same for any temperature between $-180°$ and $+450°$. This deactivation law is therefore quite characteristic and could be used to define an absolutely independent time standard.

In these experiments, it is the radioactive energy accumulated in the gas that maintains the activity of the walls. If, in fact, the gas is removed by creating a vacuum in the enclosure, it is found that the walls are then deactivated according to the rapid deactivation mode, the intensity of radiation being reduced to one-half in twenty-eight minutes. The same result is obtained when ordinary air is substituted for the active air in the enclosure.

The law of deactivation with halving in four days is therefore characteristic of the disappearance of the radioactive energy accumulated in the gas. If we use the expression adopted by Mr. Rutherford, we can say that the radium emanation disappears spontaneously as a function of time, with reduction to one-half in four days.

The emanation of thorium is of a different nature and disappears much more rapidly. The activation power is halved in about one minute ten seconds.

The emanation of actinium disappears even faster; halving occurs in seconds.

MM. Elster and Geitel have shown that there is always a very small proportion of radioactive fumes in atmospheric air, similar to the emanation emitted by radioactive bodies. Metallic wires stretched in the air and maintained at a negative potential are activated under the influence of this emanation. The air which is sucked in by means of a tube sunk into the ground is particularly charged with emanation[11]. The origin of this emanation is still unknown.

The air extracted from certain mineral waters contains emanation while the air contained in sea and river water is almost free of it.

Nature of the Emanation.

According to Mr. Rutherford, the emanation of a radioactive body is a radioactive material gas which escapes from this body. Indeed, in many ways, the emanation of radium behaves like an ordinary gas.

When two glass tanks are put into communication, one of which contains emanation while the other does not, the emanation passes by diffusing into the second tank, and when equilibrium is established, the emanation is shared between the two tanks as an ordinary gas would do: if the two tanks are at the

---

[11]Elster and Geitel, *Physik. Zeitschrift*, 15 September 1902.

same temperature, the emanation is shared between them in the ratio of their volumes; if they are at different temperatures, it is shared between them like a perfect gas obeying the laws of Mariotte and Gay-Lussac. To establish this result, it is sufficient to measure the radiation from the first reservoir before and after the partition; this radiation is proportional to the quantity of emanation contained in the tank. But, since the diffusion of the emanation requires a certain time until equilibrium is established, it is necessary, for the accuracy of the calculation relating to the experiment, to take into account the spontaneous destruction of the emanation over time[12].

The emanation of radium diffuses along a narrow tube according to the laws of gas diffusion, and its diffusion coefficient is comparable to that of carbonic acid[13].

MM. Rutherford and Soddy[14] have shown that the emanations of radium and thorium condense at the temperature of liquid air, as would gases which would be liquefiable at that temperature. A stream of emanating air loses its radioactive properties by passing through a coil which plunges into liquid air; the emanation remains condensed in the coil, and it is found in the gaseous state when it is heated. The emanation of radium condenses at $-150°$, that of thorium at a temperature between $-100°$ and $-150°$. The following experiment can be made: two closed glass vessels, one large and the other small, communicate with each other by a short tube fitted with a tap; they are filled with gas activated by radium and are therefore both luminous. We immerse the small vessel in the liquid air, all the emanation then condenses; after a while the two vessels are separated from each other by closing the

---

[12]P. Curie and J. Danne, *Comptes rendus*, 2 June 1903.
[13]P. Curie and J. Danne, *Comptes rendus*, 2 June 1903
[14]Rutherford and Soddy, *Phil. Mag.*, May 1903.

tap, and then the small vessel is removed from the liquid air. We see that it is the small reservoir that contains all the activity. To be sure, one can look at the phosphorescence of the glass in the two vessels. The large vessel is no longer bright, while the small one is brighter than at the start of the experiment. The experiment shows a particularly high brightness if care is taken to coat the walls of the two vessels with phosphorescent zinc sulfide.

However, if the emanation of radium is quite comparable to a liquefiable gas, the condensation temperature by cooling should be a function of the amount of emanation contained in a certain volume of air; this has not been reported.

It should also be noted that the emanation passes with great ease through the smallest holes or cracks in solid bodies, under conditions where ordinary material gases can only circulate with extreme slowness.

Finally, the emanation of radium differs from an ordinary material gas in that it destroys itself spontaneously when it is enclosed in a sealed glass tube; at least we observe, under these conditions, the disappearance of the radioactive property. This radioactive property is, moreover, still the only one that currently characterizes the emanation, to our knowledge, because until now we have not yet established with certainty neither the existence of a characteristic spectrum of the emanation, nor a pressure due to the emanation.

However, very recently MM. Ramsay and Soddy observed new lines in the spectrum of gases extracted from radium which, in their opinion, could belong to the emanation of radium. They also found that the gases extracted from radium contain helium, and that the latter gas forms spontaneously in the presence of the emanation of radium[15]. If these results, which are of considerable importance, are confirmed, it may be necessary to consider emanation as an unstable material gas, and helium may be one of the products of the disaggregation of this gas.

The emanations of radium and thorium do not seem to be altered by various very energetic chemical agents, and for this reason MM. Rutherford and Soddy associate them to gases of the argon family[16].

Variation in the Activity of Activated Liquids and Radium Solutions.

Any liquid becomes radioactive when it is placed in a vase in an active enclosure. If you remove the liquid from the enclosure and leave it in the open air, it rapidly deactivates by transmitting its activity to the gas and solid bodies surrounding it. If you enclose an activated liquid in a closed flask, it deactivates much more slowly and the activity then halves in four days, as would happen with an activated gas enclosed in a closed vessel. This can be explained by assuming that radioactive energy is stored in liquids in a form identical to that in which it is stored in a gas (in the form of emanation).

A solution of a radium salt behaves in a somewhat similar manner. First of all, it is remarkable that the solution of a radium salt placed for some time in a confined space is no more active than pure water placed in a vessel in the same enclosure, when the equilibrium of activity is established. If the radium solution is removed from the enclosure and left standing in the air in a wide open vessel, the activity spreads in space, and the solution

---

[15]Ramsay and Soddy, *Physikalische Zeitschrift*, 15 September 1903.
[16]*Phil. Mag.*, 1902, p. 580 ; 1903, p. 457.

becomes almost inactive, although it still contains radium. If this solution is now enclosed in a closed flask, it gradually resumes, in a fortnight, a maximum activity, which may be considerable. On the contrary, a liquid made active, but not containing radium, does not regain its activity in a closed flask after having been exposed to the open air.

Theory of Radioactivity.

The following is, according to MM. Curie and Debierne, a very general theory which makes it possible to coordinate the results of the study of induced radioactivity, results which I have just exposed and which constitute facts independent of any hypothesis[17].

It can be assumed that each atom of radium functions as a continuous and constant source of energy, without actually specifying the origin of this energy. The radioactive energy that accumulates in the radium tends to dissipate in two different ways: — (1) By radiation (rays with both charged and uncharged electricity); (2) by conduction, i.e., by gradual transmission to the surrounding bodies, by means of gases and liquids (release of an emanation and transformation into induced radioactivity).

The loss of radioactive energy, both by radiation and by conduction, increases with the amount of energy accumulated in the radioactive body. A regime of equilibrium must necessarily be established when the double loss, which I have just made mentioned, compensates for the continuous gain made by radium. This view is similar to that in use for calorific phenomena. If there is a continuous and constant release of heat in the interior of a body, the heat builds up in the body, and the temperature

---

[17]Curie and Debierne, *Comptes rendus*, 29 July 1901.

rises, until the heat loss by radiation and conduction balances the continuous supply of heat.

In general, except in certain special conditions, activity is not transmitted through solid bodies. When a solution is kept in a sealed tube, the radiation loss alone takes place, and the radiating activity of the solution is of a higher degree.

If, on the contrary, the solution is in an open vessel, the loss of activity gradually, by conduction, becomes considerable, and, when the steady state is reached, the radiating activity of the solution is very weak.

The radiant activity of a solid radium salt left in the open air does not decrease appreciably, because the propagation of radioactivity by conduction not taking place through solid bodies, it is only a very thin superficial layer that produces the induced radioactivity. It is observed, in fact, that the solution of the same salt produces much more intense induced radioactivity phenomena. With a solid salt the radioactive energy accumulates in the salt and is dissipated mainly by radiation. On the contrary, when the salt has been dissolved in water for a few days, the radioactive energy is distributed between the water and the salt, and if they are separated by distillation, the water causes a large part of the activity, and the solid salt is much less active (ten or fifteen times) than before solution. Afterwards the solid salt gradually regains its original activity.

We can further refine the above theory, supposing that the radioactivity of radium itself occurs at least largely through the radioactive energy emitted in the form of an emanation.

Each atom of radium may be considered as a constant and continuous source of emanation. At the same time as this form of energy is produced, it gradually undergoes a transformation into radioactive energy of the Becquerel radiation. The speed of this transformation is proportional to the amount of emanation accumulated.

When a radium solution is placed within an enclosure, the emanation is able to spread into the enclosure and on the walls. This is therefore what is transformed into radiation, while the solution emits only few Becquerel rays; the radiation is, in a way, externalised. On the other hand, with solid radium, the emanation not being able to escape easily, accumulates and transforms on the spot into the Becquerel radiation; this radiation therefore reaches a high value[18].

If this theory of radioactivity were general, we have to admit that all radioactive bodies give rise to an emanation. However, this emission has been observed for radium, thorium, and actinium; this latter in particular emits enormously even in the solid state. Uranium and polonium do not seem to emit any emanation, though they generate Becquerel rays. These bodies produce no induced radioactivity in an enclosed space, as do the radioactive bodies mentioned before. This fact is not in absolute contradiction with the foregoing theory. If, in fact, uranium and polonium emitted emanations which destroy themselves with very great rapidity, it would be very difficult to observe the entrainment of these emanations by air and the effects of induced radioactivity produced by them on neighbouring bodies. Such an assumption is not at all improbable, since the times required for the quantities of radium and thorium emanation to diminish to one-half are in the proportion of 5000 to 1. We shall see, moreover, that, under certain conditions, uranium can excite induced activity.

---

[18]Curie, *Comptes rendus*, 26 January 1903.

Another Form of Induced Radioactivity.

According to the law of dissipation in the open air of solid bodies activated by radium, the activity after one day is almost imperceptible.

Some bodies, however, are an exception: such are celluloid, paraffin, caoutchouc, &c. When these bodies have been activated to a sufficient degree, they lose their activity more slowly than the law can account for, and it often takes fifteen or twenty days before the activity becomes imperceptible. These bodies appear to have the property of absorbing radioactive energy in the form of an emanation; they then gradually lose it, causing induced radioactivity in their vicinity.

Induced Radioactivity with Slow Dissipation.

There is yet another very different form of induced radioactivity, which seems to occur on all bodies, when they have been kept for months in an activating enclosure. When these bodies are removed from the enclosure, their activity at first decreases to a very low value according to the ordinary law (halving in half an hour); but, when the activity has dropped to about 1/20,000 of the initial value, it does not decrease further or at least it evolves very slowly, sometimes it even increases. We have sheets of copper, aluminium, and glass which have kept a residual activity for over six months.

These phenomena of induced radioactivity seem to be of a completely different nature than ordinary ones, and they show a much slower evolution.

A considerable time is also required for both the production and dissipation of this form of induced radioactivity.

Radioactivity Induced upon Substances in Solution with Radium.

When we treat a radioactive ore containing radium, with the aim of extracting this substance, chemical separations are effected, after which the radioactivity is confined entirely to one of the products. In this way active products, which may be several hundred times as active as uranium, are separated from totally inactive products, such as copper, antimony, arsenic, &c. Certain other bodies (iron, lead) were never separated in a completely inactive state. As the active bodies concentrate, the case is no longer the same; each chemical separation no longer furnishes absolutely inactive products; all the resulting products of a separation are active in varying degrees.

After the discovery of induced radioactivity, M. Giesel was the first to try to activate ordinary inactive bismuth by keeping it in solution with very active radium. He thus obtained radioactive bismuth[19], and from this he concluded that the polonium extracted from pitchblende was probably bismuth activated by the vicinity of the radium contained in the pitchblende.

I have also prepared active bismuth by keeping bismuth in solution with a very active radium salt.

The difficulties of this experiment consist in the extreme care which must be taken to eliminate all traces of radium from the solution. If one thinks of the infinitesimal quantity of radium which is sufficient to produce in a gram of material a very notable radioactivity, it is difficult to believe in the possibility of sufficiently washing and purifying the active product. However, each purification causes a drop in activity of the activated product, whether this be due to removal of traces of radium or that the

---

[19]Giesel, *Socitè de Physique de Berlin*, January 1900.

induced radioactivity is, under these circumstances, not proof against chemical reactions.

However, the results I obtained seem to establish with certainty that activation occurs and persists after the radium has been separated. Thus by fractionating the nitrate of my activated bismuth by precipitation of the nitrogen solution by water, I find that, after very careful purification, it splits like polonium, the most active part being precipitated first.

If the purification is insufficient, the opposite occurs, indicating that traces of radium were still present with the activated bismuth. I thus obtained activated bismuth for which the direction of fractionation indicated a high purity and which was 2000 times more active than uranium. This bismuth decreases in activity over time. But another part of the same product, prepared with the same precautions and fractionating in the same manner, has retained its activity without appreciable decrease for a time which is currently around three years.

This activity is 150 times greater than that of uranium.

I have also prepared active lead and silver by leaving them in solution with radium. Generally induced radioactivity obtained in this way scarcely lessens with lapse of time, but it does not as a rule withstand many successive chemical changes of the active body.

M. Debierne[20] activated barium by placing it in solution with actinium. This barium remains active after several chemical reactions, its activity being therefore a fairly stable atomic property. Activated barium chloride fractionates like barium-radium chloride, the more active portions being the least soluble in water and dilute hydrochloric acid. The dry chloride is spontaneously lumi-

---

[20]Debierne, *Comptes rendus*, July 1900.

nous: its Becquerel radiation is similar to that of barium-radium chloride. M. Debierne has prepared an active barium chloride 1000 times as active as uranium. This barium, however, had not acquired all the characteristics of radium, for it showed none of the strongest radium lines in the spectroscope. Further, its activity diminished on standing, and after three weeks it had become one-third of its original value.

There is a wide field for research on the radioactivity induced in substances in solution with active bodies. It seems that, depending on the conditions of the experiment, it is possible to obtain more or less stable forms of atomic induced radioactivity. The radioactivity induced under these circumstances is perhaps identical with that form, which dissipates slowly, obtained by prolonged exposure at a distance in an active enclosure. We have reason to wonder to what extent atomic induced radioactivity affects the chemical nature of the atom, and whether it can modify its chemical properties, either temporarily or permanently.

The chemical studies of bodies excited at a distance is rendered difficult by the fact that the induced activity is limited to a very thin superficial layer, and that, consequently, only a very small proportion of the material has been affected.

Induced radioactivity can also be obtained by leaving certain substances in solution with uranium. The experiment is successful with barium. If, as was done by M. Debierne, sulphuric acid be added to a solution containing uranium and barium, the precipitate of barium sulphate acquires radioactivity, and, at the same time, the uranium salt loses part of its activity. M. Becquerel found that, after repeating this experiment several times, one obtains almost inactive uranium. One would think, after that, that in this operation we succeeded in separating from uranium a radioactive body different from this metal, and whose presence produced the radioactivity in uranium. This, however, is not the case, because after a few months the uranium resumes its original activity; on the contrary, the precipitated barium sulphate, loses what it had acquired.

A similar phenomenon occurs with thorium. Mr. Rutherford precipitated a solution of thorium salt with ammonia; he separated off the solution and evaporated it to dryness. He thus obtained a small very active residue, and the precipitated thorium was observed to be less active than before. This active residue, to which Mr. Rutherford gives the name of *thorium X*, loses its activity after a time, whilst the thorium regains its original activity[21].

It appears, then, that concerning induced radioactivity all bodies do not behave in a similar manner, and that certain of them are much more readily excited than others.

Dissemination of Radioactive Dust and Induced Radioactivity of the Laboratory.

When studying highly radioactive substances, special precautions must be taken if we want to be able to continue to carry out delicate measurements. The different objects used in the chemical laboratory and those used for physical experiments soon acquire radioactivity, and act upon photographic plates through black paper. Dust, room air, clothing are radioactive. The air in the room is conductive. In the laboratory where we work, the evil has reached an acute state, and we can no longer have a well-insulated device.

Special precautions should therefore be taken to avoid the spread of radioactive dust as much as possible, and to also avoid

---

[21]Rutherford and Soddy, *Zeitschr. für physik. Chemie.*, Vol. XLII, 1902, p. 81.

the phenomena of induced radioactivity.

Objects used in chemistry should never be taken into the room where physical research is carried on, and care should be taken as far as possible to allow the radioactive substances to remain unnecessarily in this room. Before starting these studies we had the habit, in the case of experiments that used static electricity, to establish a connection between the various devices by insulated metallic wires protected by metallic cylinders connected to the ground, which screened the wires against any electrical influence outside. In studies on radioactive bodies, this arrangement is absolutely defective; the air being conductive there is incomplete insulation between the wire and the cylinder, and the electromotive force of inevitable contact between the wire and the cylinder tends to produce a current through the air and to deflect the electrometer. We now put all the communication wires out of the air by placing them, for example, in the middle of cylinders filled with paraffin or some other insulating material. It would also be advantageous to use strictly closed electrometers in these studies.

Activity Induced Outside the Influence of Radioactive Substances.

Attempts have been made to produce induced radioactivity outside the action of radioactive substances.

M. Villard[22] subjected to the action of the cathode rays a piece of bismuth placed as an anticathode in a Crookes tube; this bismuth was thus made active, in fact, in an extremely weak way, because it required an exposure of eight days to obtain a photographic impression.

---

[22]Villard, *Société de Physique*, July 1900.

Mr. MacLennan has exposed different salts to the action of cathode rays, afterwards warming them slightly. The salts then acquired the property of neutralising bodies positively charged[23].

Studies of this kind are of great interest. If, using known physical agents, it were possible to create a significant radioactivity in originally inactive bodies, we might hope to find the cause of the spontaneous radioactivity of certain materials.

Variations in the Activity of Radioactive Bodies. Effects of Solution.

Polonium, as I said above, decreases in activity over time. This decrease is slow, and does not take place at the same rate with different samples. A sample of bismuth-polonium nitrate lost half of its activity in eleven months, and 95 per cent of its activity in thirty-three months. Other specimens have experienced similar diminutions.

A sample of metallic bismuth containing polonium was prepared from the nitrite, its activity after preparation being 100,000 times that of uranium. This metal is now only a body of medium radioactivity (2000 times more active than uranium). Its radioactivity is determined from time to time. In six months this metal lost 67 per cent of its activity.

The loss of activity does not seem to be facilitated by chemical reactions. In rapid chemical changes no considerable loss of activity has in general taken place.

In contrast to what happens with polonium, radium salts possess a permanent radioactivity which does not show an appreciable decrease after a few years.

---

[23]Mac Lennan, *Phil. Mag.*, February 1902.

A freshly prepared radium salt in the solid state does not at first possess a constant activity. Its activity increases from the time of preparation until it attains a practically constant limiting value after about one month. The opposite takes place for the solution. When freshly prepared, it is very active at first, but left in the open air it deactivates quick, and finally reaches a limit activity which may be considerably less than the original. These variations in activity were first observed by M. Giesel[24]. They can be very well explained from the point of view of the emanation theory. The decrease in the activity of the solution corresponds to the loss of the emanation which escapes into space; this drop is much less when the solution is contained in a sealed tube. A solution which has lost its activity in air recovers a greater activity in a sealed tube. The time of increase of the activity of the salt which, after solution, has been recently obtained in the solid state, is that during which the emanation is being newly stored in the solid radium. The following are some experiments on this subject: —

A solution of barium-radium chloride left in the air for two days becomes 300 times less active.

A solution is enclosed in a isolated vessel; we open the vessel, pour the solution into a tank, and we measure the activity: —

Activity immediately determined	67
Activity after two hours	20
Activity after two days	0.25

A solution of barium-radium chloride, which has been kept in open air is enclosed in a sealed glass tube, and the radiation from this tube is measured. The following results are found: —

---

[24]Giesel, *Wied. Ann.*, Vol. LXIX, p. 91.

Activity determined immediately	27
Activity determined after 2 days	61
Activity determined after 3 days	70
Activity determined after 4 days	81
Activity determined after 7 days	100
Activity determined after 11 days	100

The initial activity of a solid salt after preparation is lower the longer the time spent in solution. A higher proportion of the activity is then transmitted to the solvent. The following numbers give the initial activity with a chloride whose limiting activity is 800, and which were kept in solution for a given time; the salt was then dried, and its activity immediately measured: —

Limiting activity	800
Initial activity after solution and immediate evaporation	440
Initial activity after the salt has remained dissolved 5 days	120
Initial activity after the salt has remained dissolved 18 days	130
Initial activity after the salt has remained dissolved 32 days	11

During this experiment the dissolved salt was placed in a vessel merely covered with a watch-glass.

I made with the same salt two solutions which I kept in sealed tubes for thirteen months; one of these solutions was eight times more concentrated than the other: —

Initial activity of the salt in concentrated solution after drying	200
Initial activity of the salt in dilute solution after drying	100

The deactivation of the salt is therefore greater when the amount of solvent is greater, the radioactive energy transmitted to the liquid having a greater volume of liquid to saturate and a greater space to fill. The two samples of the same salt, which thus had a different initial activity, actually increased in activity at very different rates at first; after a day they had the same activity, and the increase in activity continued in exactly the same way for both of them until the limit was reached.

When the solution is dilute the loss of activity by the salt is very rapid; this is shown by the following experiments: — Three equal portions of the same radium salt are dissolved in equal amounts of water. The first solution (*a*) is left in open air for one hour, and is then evaporated. The second solution (*b*) is traversed for one hour by a stream of air, then dried. The third solution (*c*) is left exposed to the air for thirteen days, and then dried. The initial activity of each of the three salts is: —

For portion *a*   145.2
For portion *b*   141.6
For portion *c*   102.6

The limiting activity of the same salt is approximately 470. It is therefore seen that most of the effect was produced at the end of one hour. In addition, the air flow which bubbled through the solution *b* for one hour produced little effect. The proportion of salt in solution was about 0.5 per cent.

Radioactive energy in the form of an emanation is propagated with difficulty from solid radium in air; it likewise experiences resistance to propagation from solid radium in a liquid. When you stir radium sulphate with water for a whole day, its activity after this operation is practically the same as that of a portion of the same sulphate left exposed to air.

By placing the radium salt in a vacuum, all available emanation is removed. However, the radioactivity of a radium chloride kept *in vacuo* for six days was not significantly modified by this operation. This experiment shows that the radioactivity of the salt is mainly due to the radioactive energy generated within the particles, which cannot be removed by the vacuum.

The loss of activity that radium undergoes when in solution is relatively greater for the penetrating rays than for the absorbable rays. The following are examples of this: —

A radium chloride which had reached its limit of activity, 470, is dissolved and left in solution for one hour; it is then dried and its initial radioactivity is measured by the electrical method. We find that the total initial radiation is equal to the fraction 0.3 of the total limiting radiation. If the measurement of the intensity of radiation is made by covering the active body with an aluminium screen 0.01 m.m. thick, the initial radiation that passes through this screen is found to be only the fraction 0.17 of the limiting radiation traversing the same screen.

When the salt has remained thirteen days in solution, the total initial radiation is found to be the fraction 0.22 of the total limiting radiation, and is 0.13 of the limiting radiation after traversing 0.01 m.m. of aluminium.

In both cases the ratio of the initial radiation after solution to the limiting radiation is 17 times greater for the total radiation than for the radiation which passed through 0.01 m.m. of aluminium.

It should also be noted that by drying the product after dissolution, it is impossible to avoid a certain period of time during which the product is in an intermediate state, neither entirely solid nor entirely liquid. Neither can one avoid heating the product to remove the water quickly.

For these two reasons it is hardly possible to determine the true initial activity of the product which passes from the dissolved state to the solid state. In the experiments which have been just cited, equal quantities of the active bodies were dissolved in the same quantity of water, and the solutions were then evaporated to dryness under conditions as identical as possible, and without heating above 120° or 130°.

I studied the law according to which the activity of a solid radium salt increases, from the moment in which the salt is dried after dissolution until the moment in which it reaches its limiting activity. In the tables which follow, I indicate the intensity of radiation, I, as a function of time, the limiting intensity being supposed to be equal to 100, and the time being counted from the moment when the product was dried. Table I. (Fig. 13, Curve I.) refers to the total radiation. Table II. (Fig. 13, Curve II.) refers only to the penetrating rays (rays which have passed through 3 c.m. of air and 0.01 m.m. of aluminium).

Table I

Time. Days.	I.
0	21
1	25
3	44
5	60
10	78
19	93
33	100
63	100

Table II

Time. Days.	I.
0	1.3
1	19
3	43
6	60
15	70
23	86
46	94

I have made several other series of measurements of the same kind, but they are not absolutely in agreement with each other, although the general character of the curves obtained remains the same. It is hard to obtain consistent results. It can be noted, however, that the resumption of activity requires more than one month for its production, and that the most penetrating rays are the most deeply affected by the effect of dissolution.

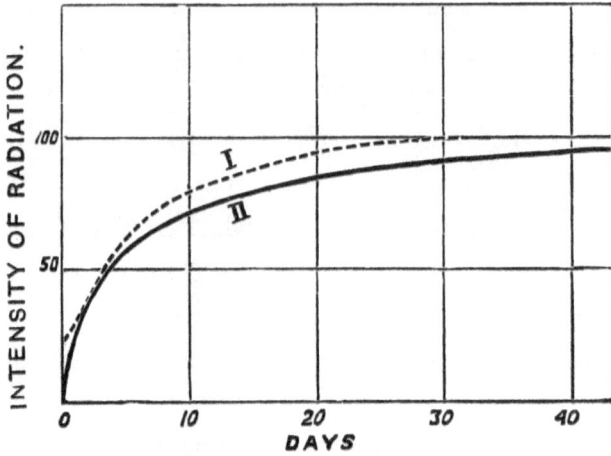

Figure 13.

The initial intensity of the radiation which can pass through 3 c.m. of air and 0.01 m.m. of aluminium is only 1 per cent of the limiting intensity, whilst the initial intensity of the total radiation is 21 per cent of the total limiting radiation.

A radium salt which has been dissolved and has just been dried, has the same power to cause induced activity (and, consequently, of allowing the escape of an emanation), as a sample of the same salt which, after having been prepared in the solid state, has remained in this state long enough to have reached its limiting radioactivity. The radiant activity of these two products is, however, extremely different; the first is, for example, five times less active than the second.

Variations of the Activity of Radium Salts on Heating.

When a radium compound is heated, it gives off an emanation and loses activity. The loss of activity is all the greater as the heating is both more intense and more prolonged. Thus, on heating a radium salt for one hour to 130°, it loses 10 per cent of its total radiation; on the contrary, heating for ten minutes to 400° produces no apparent effect. Heating to redness for several hours destroys 77 per cent of the total radiation.

The loss of activity by heating is greater for penetrating rays than for absorbable rays. Thus heating for a few hours destroys approximately 77 percent of the total radiation, but the same heat destroys almost all (99 percent) of the radiation which is able to pass through 3 c.m. of air and 0.01 m.m. of aluminum. By keeping the barium-radium chloride fused for a few hours (around 800°), 98 percent of the radiation capable of passing through 0.3 m.m. of aluminum is destroyed. We can say that the penetrating rays do not exist appreciably after a strong and prolonged heating.

When a radium salt has lost part of its activity by heating, this drop in activity does not persist: the activity of the salt regenerates spontaneously at room temperature and tends towards a certain limit value. I observed the curious fact that this limit is higher than the limit activity of the salt before heating, at least this is the case for chloride. Here are some examples: a sample of barium-radium chloride which, after having been prepared in the solid state, has long since reached its limit activity, has a total radiation represented by the number 470, and a radiation capable of crossing 0.01 m.m. of aluminum, represented by the number 157. This sample is subjected to heating to redness for a few hours. Two months after heating, it reaches a limit activity with a total radiation equal to 690, and a radiation through 0.01 m.m. of aluminum equal to 227. The total radiation and the radiation which crosses aluminum are therefore increased respectively in ratios $\frac{690}{470}$ and $\frac{270}{156}$ — These two ratios are practically equal to one another, and are equivalent to 1.45.

A sample of radium-barium chloride which, after having been prepared in the solid state, has reached a limiting activity of 62, is kept fused for some hours; the molten product is then powdered. The product regains a new limiting activity equal to 140, which is twice as great as that to which it was able to attain when prepared in the solid state without having been sensibly heated during evaporation.

I studied the law of increasing the activity of radium compounds after heating. The following are the results of two series of determinations: — The numbers of Table I. and II. represent the intensity of the radiation (I) as a function of time, the limiting intensity being assumed to be equal to 100, and the time being counted from the end of heating. Table I. (Fig. 14, Curve I.) refers to the total radiation of a sample of barium-radium chloride. Table II. (Fig. 14, Curve II.) relates to the penetrating radiation of a sample of barium-radium sulphate, the intensity of the radiation which traversed 3 c.m. of air and 0.01 m.m. of aluminium having been measured. The two products were heated to a cherry red for seven hours.

I made several other series of determinations, but the results did not agree well.

The effect of heating does not persist when the heated radiant substance is dissolved. Of two samples of the same radiant substance of activity 1800, one was strongly heated, and its activity was reduced by heating to 670. The two samples having been at this time dissolved and left to dissolve for 20 hours, their initial solid state activities were 460 for the unheated product and 420 for the heated product; there was therefore no significant difference between the activity of these two products. However, if the two products do not remain in solution for a sufficient time, if, for example, they are dried immediately after having dissolved

Table I

Time. Days.	I.
0	16.2
0.6	25.4
1	27.4
2	38
3	46.3
4	54
6	67.5
10	84
24	95
57	100

Table II

Time. Days.	I.
0	0.8
0.7	13
1	18
1.9	26.4
6	46.2
10	55.5
14	64
18	71.8
27	81
36	91
50	95.5
57	99
84	100

them, the unheated product is much more active than the heated product; it takes a while for the dissolving state to dissipate the heating effect. A product of activity 3200 was heated and only had an activity of 1450 after heating. This product was dissolved at the same time as a portion of the same unheated product, and the two portions were dried immediately. The initial activity was 1450 for the unheated product and 760 for the heated one.

For solid radium salts, the power to cause the induced radioactivity is strongly influenced by heating. While the radiant compounds are heated, they give off more emanation than at room temperature; but, when they are then brought back to room temperature, not only their radioactivity is much lower than that which they had before heating, but also their activating power is considerably reduced. During the time following heating, the radioactivity of the product increases and may even exceed the initial value. The activating power is also partially restored; how-

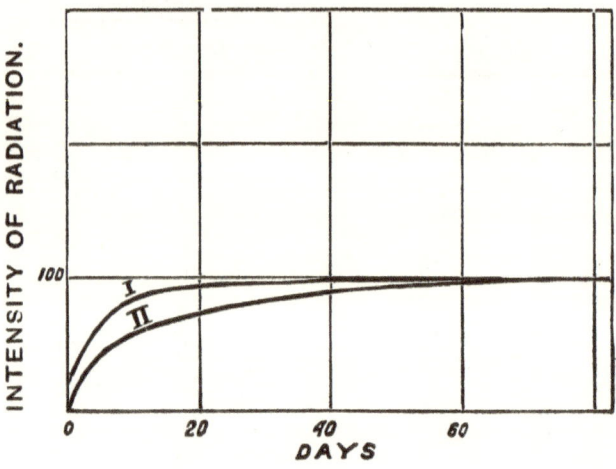

Figure 14.

ever, after prolonged heating to redness, almost all of the activating power is removed, without spontaneous re-appearance
afterwards. The radiant salt can be restored to its original activating power by dissolving it in water and drying it in an oven
at a temperature of 120°. It therefore seems that the calcination
has the effect of putting the salt in a particular physical state, in
which the emanation is released with much more difficulty than
for the same solid product which has not been heated to high
temperature, and it follows quite naturally that the salt reaches a
higher radioactivity limit than that which it had before the heating. To restore the salt to the physical state it had before heating,
simply dissolve it and dry it, without heating it, above 150°.

The following are numerical examples of the above:—

I represent by a the limit of induced activity produced in a
closed vessel on a copper plate by a sample of barium-radium
carbonate of activity 1600.

Let us assume $a = 100$ for the not heated product. We find —

1 day after heating $a = $ 3.3
4 days after heating $a = $ 7.1
10 days after heating $a = 15$
20 days after heating $a = 15$
37 days after heating $a = 15$

The radioactivity of the product had decreased by 90 per cent by heating, but after one month it had already returned to its original value.

Here is a similar experiment made with a barium-radium chloride of activity 3000. The induction capacity is determined in the same manner as before.

For the product not heated $a = 100$.

Induction capacity of the product after being heated to redness for three hours: —

2 days after heating	2.3
5 days after heating	7.0
11 days after heating	8.2
18 days after heating	8.2
Activating power of unheated product which has been dissolved and then dried at 150°	92
Activating power of the heated product which has been dissolved and then dried at 150°	105

Theoretical Interpretation of the Causes of the Variations in Activity of Radium Salts after Solution and after Heating.

The facts which have just been exposed can be, in part, explained by the theory according to which radium produces energy in the form of emanation, the latter then being transformed into radiant energy. When a radium salt is dissolved, the emanation it produces spreads outside the solution and causes radioactivity outside the source from which it comes; when the solution is evaporated, the solid salt obtained is not very active, since it contains only little emanation. Gradually the emanation is accumulated in the salt, the activity of which rises to a limiting value, which is reached when the production of the emanation by the radium compensates the loss by external emission and by local transformation into Becquerel rays.

When a radium salt is heated, the emanation rate outside the salt is greatly increased, and the phenomena of induced radioactivity are more intense than when the salt is at room temperature. But when the salt returns to room temperature, it is exhausted, as in the case where it had been dissolved, it contains little emanation, its activity having become greatly reduced. Gradually the emanation accumulates anew in the solid salt, and the radiation increases.

It can be assumed that radium gives rise to a constant flow of emanation — part of which escapes outside, the remainder being transformed in the radium itself into Becquerel rays. When radium is heated to redness, it loses most of its capacity to activate by induction; in other words, the evolution of the emanation is reduced. Consequently, the proportion of the emanation used in the radium itself should be higher, and the substance attains a higher limit of radioactivity.

We can propose to theoretically establish the law of the increase of the activity of a solid radiferous salt which has been dissolved or which has been heated. We will assume that the intensity of the radium radiation is, at all times, proportional to the quantity of emanation q present in the radium. We know that the emanation destroys itself spontaneously according to a law such that we have, at all times —

$$q = q_0 e^{-\frac{t}{\theta}} \tag{1}$$

q_0 being the amount of the emanation at the beginning of the observation, and θ the time constant, equal to 4.97×10^5 secs.

Now let Δ be the evolution of the emanation by radium, a quantity which I will assume constant. Let us consider what would occur if no emanation were escaping to the exterior. The emanation generated would then be completely utilised by the radium for the production of the radiation. We have from Formula 1 —

$$\frac{dq}{dt} = -\frac{q_0}{\theta}e^{-\frac{t}{\theta}} = -\frac{q}{\theta};$$

and consequently, in the state of equilibrium, the radium would contain a certain quantity of emanation, Q, such that —

$$\Delta = \frac{Q}{\theta} \tag{2}$$

and the radiation of the radium would then be proportional to Q.

Suppose we put radium in conditions where it loses the emanation to the exterior; this can be achieved by dissolving or heating the radiant compound. The balance will be disturbed and the activity of radium will decrease. But as soon as the cause of the loss of emanation has been removed (the body has returned to the solid state, or else it has stopped heating), the emanation accumulates again in the radium, and we have a period, during which the flow Δ prevails over the speed of destruction, $\frac{q}{\theta}$. We then have —

$$\frac{dq}{dt} = \Delta - \frac{q}{\theta} = \frac{Q - q}{\theta},$$

from which —

$$\frac{d}{dt}(Q-q) = -\frac{Q-q}{\theta},$$

$$Q - q = (Q - q_0)e^{-\frac{t}{\theta}} \qquad (3)$$

q_0 being the amount of emanation present in the radium at time $t = 0$.

According to Formula 3, the excess of the quantity of emanation, Q, contained by the radium in a state of equilibrium above the quantity, q contained at a given moment, decreases as a function of time according to an exponential law, which is also the law of the spontaneous disappearance of the emanation. The radiation of radium being proportional to the amount of emanation, the excess of the intensity of the limiting radiation above the actual intensity should decrease as a function of the time by the same law; this excess must thus be halved in about four days.

The above theory is incomplete, since the loss of emanation to the exterior has been ignored. It is, moreover, difficult to know how it evolves as a function of time. By comparing the results of the experiment with those of this incomplete theory, one does not find a satisfactory agreement; however, we retain the conviction that the theory in question contains some truth. The law according to which the excess of the limit activity above the actual activity decreases by half in 4 days represents, with a certain approximation, the progress of the resumption of activity after heating for ten days. In the case of resumption of activity after dissolution this same law seems to be more or less suitable for a certain period of time, which begins two or three days after the drying of the product and continues for ten to fifteen days. The phenomena are moreover complex; the theory indicated does not explain why the penetrating rays are suppressed in a higher proportion than the absorbable rays.

Nature and Cause of Radioactivity Phenomena.

From the beginning of research on radioactive bodies, and while the properties of these bodies were still barely known, the spontaneity of their radiation presented itself as a problem having the greatest interest for physicists. Today we are more advanced in the understanding of radioactive bodies, and we know how to isolate a very powerful radioactive body, viz., radium. The use of the remarkable properties of radium has made it possible to make an in-depth study of the rays emitted by radioactive bodies; the various groups of rays under investigation present analogies with the groups of rays existing in Crookes tubes: cathode rays, Röntgen rays, canal rays. The same groups of rays are also found in the secondary radiation produced by Röntgen rays[25], and in the radiation of bodies which have acquired the induced radioactivity.

But if the nature of the radiation is actually better known, the cause of this spontaneous radioactivity remains mysterious and this phenomenon is always an enigma to us and a subject of deep astonishment.

The spontaneously radioactive bodies, primarily radium, are sources of energy. The energy flow to which they give rise is revealed to us by Becquerel radiation, by chemical and luminous effects, and by the continuous release of heat.

It has often been asked whether energy is created in radioactive bodies themselves or whether it is borrowed by these bodies

[25]Sagnac, *Thèse de doctorat.* — Curie and Sagnac, *Comptes rendus*, April 1900.

from external sources. None of the many hypotheses, which result from these two possibilities, have yet received experimental confirmation.

Presumably, the radioactive energy has been stored previously and is gradually depleting as it happens with very long-lived phosphorescence. One can imagine that the release of radioactive energy corresponds to a transformation of the very nature of the atom of the radiant body which is in the process of evolution; the fact that radium releases heat continuously argues in favor of this hypothesis. We can assume that the transformation is accompanied by a loss of weight and an emission of material particles which constitute the radiation. The source of energy can still be sought in the energy of gravitation. Finally, one can imagine that space is constantly crossed by radiations still unknown which are stopped in their passage through radioactive bodies and transformed into radioactive energy.

There are many reasons to be put forward for and against these various ways of seeing things, and more often than not tests of experimental verification of the consequences of these hypotheses have given negative results. The radioactive energy of uranium and radium apparently neither becomes exhausted nor varies appreciably over time. Demarçay examined spectroscopically a sample of pure radium chloride after a five months' interval, and observed no change in the spectrum. The principal barium line, which was visible in the spectrum indicating the presence of a trace of barium, had not increased in intensity during the interval, showing therefore that there was no transformation of radium into barium to an appreciable extent.

The variations of weight announced by M. Heydweiller in

radium compounds[26] cannot yet be considered as established facts.

Elster and Geitel found that the radioactivity of uranium is not affected at the bottom of a mine-shaft 850 m. deep; a layer of earth of this thickness would therefore not affect the hypothetical primary radiation which would be excited by the radioactivity of uranium.

We have determined the radioactivity of uranium at noon and at midnight, thinking that if the hypothetical primary radiation had its origin in the sun it would be partly absorbed by crossing the earth. The experiment showed no difference in the two measurements.

The most recent research supports the hypothesis of an atomic transformation of radium. This hypothesis was presented at the start of the research on radioactivity[27]; it was openly adopted by Mr. Rutherford who assumed that the emanation of radium is a material gas which is one of the products of the disintegration of the atom of radium[28]. The recent experiments of MM. Ramsay and Soddy tend to prove that emanation is an unstable gas which is destroyed by producing helium. On the other hand, the continuous flow of heat provided by radium cannot be explained by an ordinary chemical reaction, but could perhaps have its origin in a transformation of the atom.

Finally, it should be remembered that new radioactive substances are still found in uranium ores; we have searched in vain for radium in the commercial barium. The presence of radium therefore seems to be linked to that of uranium. Urane ores also contain argon and helium, and this coincidence is probably not

---

[26]Heydweiller, *Physik. Zeitschr.*, October 1902.
[27]Mme Curie, *Revue générale des Sciences*, 30 January 1899.
[28]Rutherford and Soddy, *Phil. Mag.*, May 1903.

due to chance either. The simultaneous presence of these various bodies in the same ores suggests that the presence of some is perhaps necessary for the formation of others.

It should be noted, however, that the facts which support the idea of an atomic transformation of radium may also be given a different interpretation. Instead of assuming that the radium atom is transformed, one could assume that this atom is itself stable, but that it acts on the medium which surrounds it (neighboring material atoms or vacuum ether) so as to give rise to atomic transformations. This hypothesis leads just as well to admitting the possibility of the transformation of the elements, but the radium itself would then no longer be an element in the process of disintegration.

Index